The Open University

S342

Science: a third level course

PHYSICAL CHEMISTRY

PRINCIPLES OF CHEMICAL CHANGE

BLOCK 8
DYNAMIC ELECTROCHEMISTRY

THE S342 COURSE TEAM

CHAIR AND GENERAL EDITOR

Kiki Warr

AUTHORS

Keith Bolton (Block 8; Topic Study 3)
Angela Chapman (Block 4)
Eleanor Crabb (Block 5; Topic Study 2)
Charlie Harding (Block 6; Topic Study 2)
Clive McKee (Block 6)
Michael Mortimer (Blocks 2, 3 and 5)
Kiki Warr (Blocks 1, 4, 7 and 8; Topic Study 1)
Ruth Williams (Block 3)

Other authors whose previous S342 contribution has been of considerable value in the preparation of this Course

Lesley Smart (Block 6)
Peter Taylor (Blocks 3 and 4)
Dr J. M. West (University of Sheffield, Topic Study 3)

COURSE MANAGER

Mike Bullivant

EDITORS

Ian Nuttall
Dick Sharp

BBC

David Jackson
Ian Thomas

GRAPHIC DESIGN

Debbie Crouch (Designer)
Howard Taylor (Graphic Artist)
Andrew Whitehead (Graphic Artist)

COURSE READER

Dr Clive McKee

COURSE ASSESSOR

Professor P. G. Ashmore (original course)
Dr David Whan (revised course)

SECRETARIAL SUPPORT

Debbie Gingell (Course Secretary)
Jenny Burrage
Margaret Careford

The Open University, Walton Hall, Milton Keynes, MK7 6AA

Edited, designed and typeset by The Open University.

Printed in the United Kingdom by Henry Ling Ltd, The Dorset Press, Dorchester DT1 1HD

ISBN 0 7492 51891

This text forms part of an Open University Third Level Course. If you would like a copy of Studying with The Open University, please write to the Central Enquiry Service, PO Box 200, The Open University, Walton Hall, Milton Keynes, MK7 6YZ. If you have not enrolled on the Course and would like to buy this or other Open University material, please write to Open University Educational Enterprises Ltd, 12 Cofferidge Close, Stony Stratford, Milton Keynes, MK11 1BY, United Kingdom.

s342block8i1.2

CONTENTS

1 INTRODUCTION 5

2 THE ELECTRICAL POTENTIAL DIFFERENCE 6

2.1 How a potential difference arises 6
2.2 What value does the potential difference take? 7
2.3 Variation of potential with distance 11
2.4 Summary of Section 2 11

3 THE EFFECT OF THE POTENTIAL DIFFERENCE ON THE RATE OF REACTION 12

3.1 Experimental behaviour 12
3.2 A simple qualitative approach to reaction rates 13
3.3 The Butler–Volmer equation 17
3.4 Summary of Section 3 20

4 OBTAINING EXPERIMENTAL DATA 21

5 USING TAFEL DATA 23

5.1 i_e values 23
5.2 a values 24
5.3 Tafel plots and metal extraction 26
5.4 Summary of Section 5 31

6 THE VARIATION OF CELL POTENTIAL WITH CURRENT 31

6.1 Self-driving cells 32
6.2 Driven cells 36
6.3 The general relationship between V and I 37
6.4 Summary of Section 6 38

7 INDUSTRIAL ELECTROLYTIC PROCESSES 40

7.1 Introduction 40
7.2 The chlor-alkali industry: an example of ‘electrochemistry in action’ 40
7.3 Electrosynthesis: a few general remarks 48

8 BATTERIES AND OTHER ELECTROCHEMICAL POWER SOURCES 51

8.1 Fuel cells 54

APPENDIX 1 56

OBJECTIVES FOR BLOCK 8 57

SAQ ANSWERS AND COMMENTS 59

ANSWERS TO EXERCISES 65

ACKNOWLEDGEMENTS 69

1 INTRODUCTION

Block 7 considered in detail the thermodynamic aspects of electrochemical reactions, but we know from our earlier studies of chemistry that to consider thermodynamic aspects alone is not enough; we must also consider the kinetic factors affecting reactions, that is, the rates of the reactions. This was highlighted in the final Section of Block 7. Zinc metal is extracted electrolytically from an aqueous solution using a potential difference of around 3.5 V. Thermodynamics tells us that at this potential difference the process is indeed possible, but so is the electrolysis of water, so we could expect both evolution of hydrogen and deposition of zinc at the cathode. However, evolution of hydrogen doesn't occur in practice, so this alternative reaction is not a problem. Why? The explanation given is that although the production of hydrogen is thermodynamically favourable under these conditions, the *rate* of its production is low. So in order to predict the course of electrochemical reactions we must, not surprisingly, consider the kinetic as well as the thermodynamic aspects.

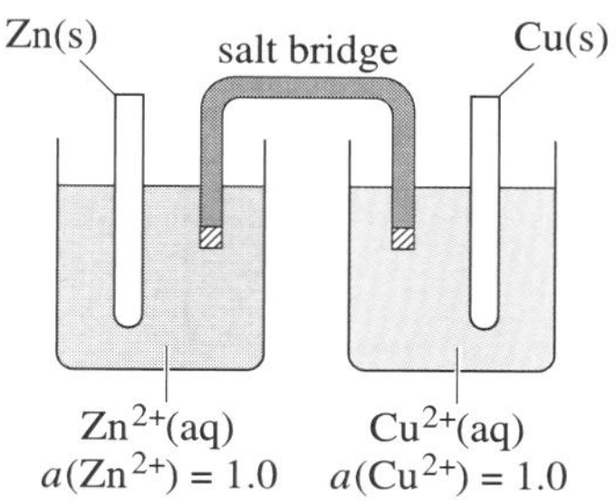

Figure 1 The Daniell cell, set up under standard conditions.

Consider the electrochemical cell shown in Figure 1. There are many questions that we could ask about this cell that are relevant to a kinetic study of the system, for example:

1 What reactions are likely at the left-hand electrode?

From your knowledge so far, you would probably answer that the reaction is

$$Zn(s) = Zn^{2+}(aq) + 2e \qquad (1)$$

and indeed this answer is sufficient for a thermodynamic study. However, for a kinetic discussion we need to get involved in the details of the mechanism of the reaction; we need information on what steps are involved, and on the step that is likely to be rate-limiting.

2 Where is reaction taking place?

The likely response to this question is 'at the electrode'. However, we shall discover that reaction takes place not on the surface of the electrode but at a very short distance from it.

3 How does a potential difference arise?

We know that the potential difference of the cell shown in Figure 1 is 1.10 V under 'imposed' equilibrium conditions, that is, when no current is being drawn from the cell, but we haven't yet discussed how such a potential difference arises.

4 What affects the rate of an electrochemical reaction?

The rate depends, as you might expect from your knowledge of solution chemistry, on the concentrations of the various species and the temperature of the system, but are there any other important variables to consider?

These are just a few of the many questions that must be answered before we can begin to understand the kinetics of electrode processes. For the first half of this Block (up to the end of Section 5) we shall restrict our discussion to the processes occurring at *one electrode*. This will allow us to develop many of the concepts required to understand the rates of electrochemical processes, while keeping discussion relatively simple. Detailed discussion of what happens when two electrodes are brought together to form a 'real' electrochemical device is left until the second half of the Block. Please remember this point; if you do, it will greatly aid understanding of some of the concepts to be presented.

Before we embark on a kinetic study of electrochemical reactions, we need to develop a deeper understanding of potential difference; how it arises, what value it takes, over what distance it operates and whether or not it is likely to influence the rate of reaction.

STUDY COMMENT There are no video or AV sequences associated with this Block, but you should note that the video sequence *Fighting rust in your car* (band 9 on videocassette 2) acts as a 'leader' to Topic Study 3.

2 THE ELECTRICAL POTENTIAL DIFFERENCE

2.1 How a potential difference arises

This is one area where it is much easier to consider the processes at a single electrode before discussing a complete electrochemical cell.

Consider the construction of the zinc/zinc ion electrode which forms part of the cell shown in Figure 1. A piece of zinc metal containing no excess charge (and so assumed to be at zero electrical potential) is immersed in a solution containing zinc ions, in which the amount of positive charge is balanced by the amount of negative charge (and so is also assumed to be at zero electrical potential)*. At the instant of immersion, the electrical potential difference between the metal and the solution is zero. This picture then changes rapidly. The following reaction starts to take place in the forward direction:

$$Zn(s) = Zn^{2+}(aq) + 2e \qquad (1)$$

that is, zinc atoms break away from the crystal lattice and enter the solution as positive ions. This leaves behind electrons, which reside on the metal surface. The solution thus gains a net positive charge and the metal a net negative charge. Thus, even though the amount of net positive charge in the solution is balanced by the amount of net negative charge on the metal, such that the total system, solution plus metal, is electrically neutral, this charge transfer reaction has led to a **charge separation** between the two phases. This charge separation means that an *electrical potential difference* operates between the two phases. It is usual to refer to this term as the **potential difference**, as we did in Block 7; we shall give it the symbol $\Delta\phi$.

The excess electrons on the metal and the excess positive ions in the solution are not free to wander around their respective phases, because there will be an electrostatic attraction between them. According to the simplest model, the result is that the excess of positively charged ions line up alongside the excess of electrons in the surface layer of the metal: the excess positive charge of the solution is balanced by the excess negative charge on the metal. This distribution of charge is often referred to as the **electrical double layer** (Figure 2). With the excess charge being confined to this double layer, we can expect the potential difference to be confined to a distance very close to the electrode surface. The region of interest is therefore this phase boundary, or *interface*.

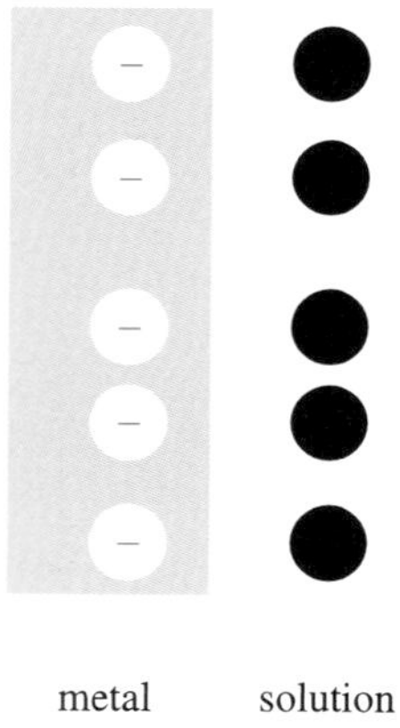

Figure 2 Representation of the electrical double layer at a metal–solution interface.

With charge transfer reactions like the one in equation 1 it is easy to see how a potential difference could develop across an interface. However, interfacial potential differences are also generated by other processes. Because the forces between species at an interface are different from those that exist within the bulk of the medium, a potential difference tends to be generated across the interface, without the necessity for charge transfer. Thus, in the construction of a zinc/zinc ion half-cell, a potential difference will arise at the metal–solution interface through processes other than charge transfer reactions. Similarly, a potential difference will exist between the solution and the walls of the container shown in Figure 1. Moreover, if the zinc electrode is connected to an external circuit, by a copper wire, say, there will also be a potential difference between the zinc and copper metals. Thus,

Potential differences are a property of all interfaces.

Here we shall not go into the details of how such potential differences arise, but you should be aware that they exist. Suffice it to say that the potential differences arising from processes other than charge transfer reactions are often small; they are sometimes neglected altogether.

* These assumptions relating zero excess charge to zero potential difference are strictly valid only if there is no interaction between the phase under consideration and its environment.

2.2 What value does the potential difference take?

Dipping a metal into a solution of its ions produces a potential difference across the metal–solution interface. At the instant of immersion of the metal, the potential difference is zero, but after a period of time (which is usually short) a potential difference develops. But what is the magnitude of this potential difference? For example, is there any limit to the amount of charge transfer that can take place?

To examine this question, consider again the charge transfer process in reaction 1:

$$Zn(s) = Zn^{2+}(aq) + 2e \qquad (1)$$

As the oxidation reaction takes place, the electrode becomes more negatively charged and the solution side of the interface more positively charged; evidently the magnitude of the potential difference must increase.

- Is this going to affect the rate of the reaction?
- Yes. The release of more electrons in a negatively charged environment (or the formation of more positive ions in a positively charged environment) is going to become more difficult as the reaction proceeds.

> The rate of the reaction is affected by the potential difference across the interface.

This effectively answers Question 4 in Section 1: in electrochemical reactions the other important variable to consider, besides temperature and concentration factors, is the potential difference across the interface.

A change in the potential difference can have a dramatic effect on the rate of an electrochemical reaction, dwarfing any effects from modest temperature or concentration changes. It is *the* most important variable in electrochemical reactions. As you will see later on, a change in the potential difference by as little as 1 V can change the rate of reaction by a factor of 10^9.

awesome!!!

But to return to reaction 1: as the potential difference increases, so the rate of the oxidation reaction 1 will decrease. By the same token, we would expect the rate of the *reverse* of reaction 1, namely the reduction reaction, to become significant as the electrode becomes more negatively charged: zinc ions will gain electrons from the zinc metal surface and be deposited as zinc atoms. This in turn suggests that, as with all chemical reactions, an equilibrium will eventually be established, in which the rate of the forward, oxidation, reaction equals the rate of the reverse, reduction, reaction. So the potential difference, which initially was zero, will increase and will continue to increase (by virtue of charge transfer processes) until a particular constant value is reached. This value, plus (usually smaller) contributions arising from other processes, not involving electron transfer, is called the **equilibrium potential difference**, and is given the symbol $\Delta\phi_e$. Obviously the actual value depends on the interface being studied, its temperature and the concentrations of the various species present.

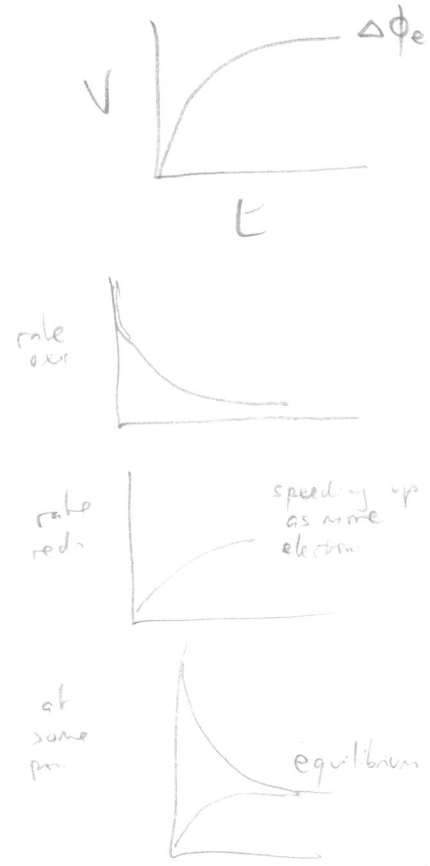

Try not to be confused by the above terminology. Remember, we are referring here to the processes taking place at a *single* electrode. In general, at a given electrode at any particular time, *both* oxidation and reduction processes are taking place. In studying Block 7, however, you became used to referring to oxidation at one electrode (the anode) and reduction at the other electrode (the cathode). But what we really mean when we talk in such terms is that there is *net* oxidation at one electrode and *net* reduction at the other; both oxidation *and* reduction reactions will be going on at each electrode. However, if the rate of the oxidation reaction at a given electrode exceeds that of the corresponding reduction reaction, then *net* oxidation takes place at that electrode. This and other situations for the zinc/zinc ion electrode are shown in Figure 3. Until this point, we haven't been concerned with the competing processes at a single electrode, but now they are of vital importance to our understanding of the kinetics of electrode processes.

$$Zn \longrightarrow Zn^{2+} + 2e$$
$$Zn \longleftarrow Zn^{2+} + 2e \qquad \text{net oxidation}$$

$$Zn \longrightarrow Zn^{2+} + 2e$$
$$Zn \longleftarrow Zn^{2+} + 2e \qquad \text{equilibrium}$$

$$Zn \longrightarrow Zn^{2+} + 2e$$
$$Zn \longleftarrow Zn^{2+} + 2e \qquad \text{net reduction}$$

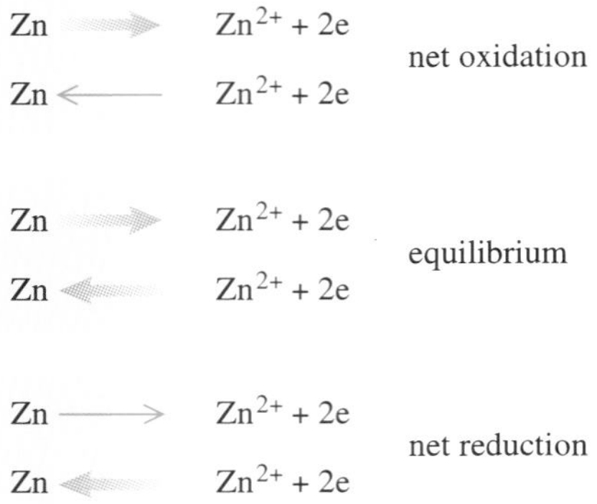

Figure 3 Representation of the various processes possible at a zinc/zinc ion electrode. The thick arrows represent more reaction than the thin arrows.

It would be interesting to compare values of the equilibrium potential difference for different electrode systems, but how do we determine such values? Well, we could imagine setting up a system to measure the potential difference across a zinc/zinc ion electrode as shown in Figure 4, for example. But if we allow equilibrium to be established, would the potential difference recorded by the voltmeter, that is V, give the value of $\Delta\phi_e$ across the interface between the zinc electrode and the zinc ion solution? Unfortunately, the answer is no. The recorded potential difference involves several other interfaces as well, and there could be a contributory potential difference across each of them.

- Study Figure 4, and identify the interfaces involved (you do not need to consider the connections to, or within, the digital voltmeter – such potential differences can be regarded as negligible.)

- The above measurement involves the following interfaces:

    Pt wire|Zn electrode|aqueous zinc solution|Pt wire

or, written in the shorthand familiar from Block 7,

    Pt(s)|Zn(s)|Zn²⁺(aq)|Pt(s)

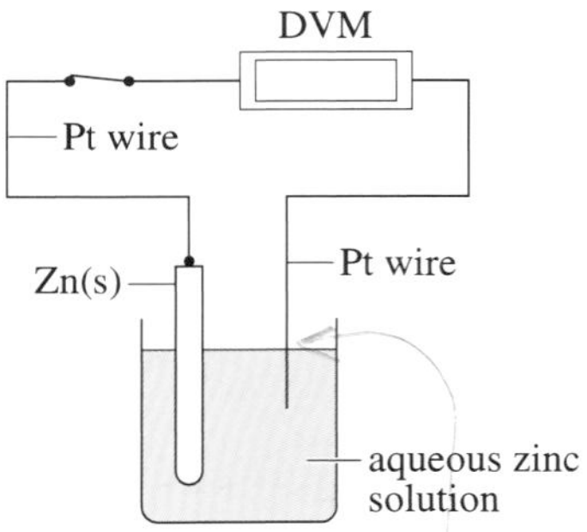

Figure 4 A system set up to attempt to measure the potential difference between zinc metal and aqueous zinc ions.

The expression for V can be derived quite simply by representing the interfaces as above, and then writing the potential difference at each interface, and adding the terms as we move *from the right to the left*. This may seem a little odd, but it is consistent with the sign conventions that we met in Block 7, as we shall see in a moment.

Moving from the right, the first interface is that between Pt wire and the aqueous zinc solution. We can represent the potential difference across this interface as $^{Pt}\Delta^{Zn^{2+}}\phi$, where

$$^{Pt}\Delta^{Zn^{2+}}\phi = \phi(Pt, s) - \phi(Zn^{2+}, aq) \tag{2}$$

If we continue this for all the interfaces, then:

$$V = {}^{Pt,RHE}\Delta^{Zn^{2+}}\phi + {}^{Zn^{2+}}\Delta^{Zn}\phi + {}^{Zn}\Delta^{Pt,LHE}\phi \tag{3}$$

where we have used 'RHE' and 'LHE' to distinguish the Pt wire on the right of the cell from that on the left.

At equilibrium (i.e. when no current flows in the external circuit), we can write

$$V_e = {}^{Pt,RHE}\Delta^{Zn^{2+}}\phi_e + {}^{Zn^{2+}}\Delta^{Zn}\phi_e + {}^{Zn}\Delta^{Pt,LHE}\phi_e \tag{4}$$

where $V_e = E$, the emf of the cell.

Rewriting this expression in terms of the potentials

$$V_e = \{\phi_e(Pt, RHE) - \phi_e(Zn^{2+}, aq)\} + \{(\phi_e(Zn^{2+}, aq) - \phi_e(Zn, s)\} + \{\phi_e(Zn, s) - \phi_e(Pt, LHE)\}$$

most of the terms cancel out, leaving $V_e = \phi_e(Pt, RHE) - \phi_e(Pt, LHE)$, which is the convention adopted in Block 7 for recording the emf of a cell.

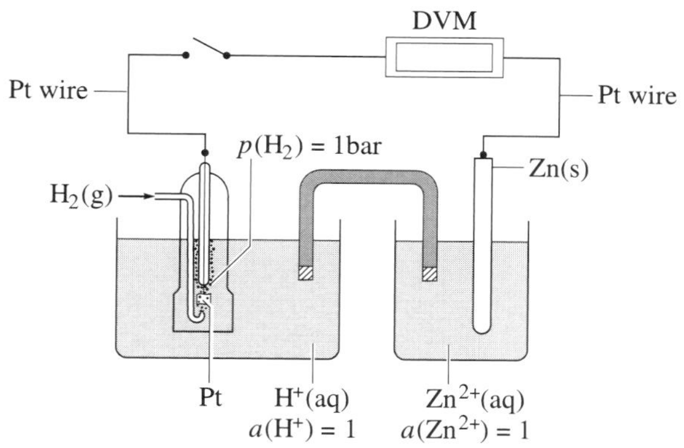

Figure 5 Measurement of the standard electrode potential for the Zn^{2+}|Zn system.

More importantly, it is impossible to extract the value of $^{Zn^{2+}}\Delta^{Zn}\phi_e$, the quantity we are interested in, from equation 4. This situation arises *whenever* we try to measure a value for the potential difference across *one* interface, because any measurable potential difference always involves several different interfaces.

But, *it is the potential difference across the metal–solution interface that determines the rate of an electrochemical reaction*, and it seems that to make progress we shall have to get at its value in some way. So how do we proceed? We shall show that we can relate reaction rates to *changes* in the potential difference across a particular interface. Such changes in potential difference can be measured experimentally.

Before leaving this point, you may well be wondering what is the distinction between the equilibrium potential difference across a metal–solution interface, $\Delta\phi_e$, and the *electrode potential*, E, of that system. Some of the salient points are brought out by referring back to Block 7 and considering how electrode potential values are determined. For example, the *standard* electrode potential of the Zn^{2+}|Zn system can be measured using the set-up shown in Figure 5.

- Examine Figure 5 and write down the interfaces involved (again you can neglect any connections within the DVM; you can also neglect any interfaces associated with the salt bridge.)

- The interfaces involved are:

  $Pt(s)$|$Pt,H_2(g)$|$H^+(aq)$|$Zn^{2+}(aq)$|$Zn(s)$|$Pt(s)$

You may well be a bit puzzled at this point. When we introduced the notation for writing cell diagrams in Block 7, we did not include the connecting leads – the platinum wires in Figure 5. We have done so now in order to highlight the fact that the measured cell potential, V, includes contributions from the potential difference across *each* of the interfaces identified above. By convention, all electrode potential values are determined relative to the standard hydrogen electrode *using platinum connecting leads*, as shown in Figure 5.

- Now write down an expression for the potential difference, V, for the cell with no current being drawn. Remember to start from the right-hand side.

- If no current is being drawn from the cell the potential differences will be the equilibrium potential differences, in which case $V = V_e = E$. Therefore,

  $E = {}^{Pt}\Delta^{Zn}\phi_e + {}^{Zn}\Delta^{Zn^{2+}}\phi_e + {}^{Zn^{2+}}\Delta^{H^+}\phi_e + {}^{H^+}\Delta^{H_2,Pt}\phi_e + {}^{Pt,H_2}\Delta^{Pt}\phi_e$

If the salt bridge is working effectively, $^{Zn^{2+}}\Delta^{H^+}\phi_e$ can be put as zero. Thus,

  $E = {}^{Pt}\Delta^{Zn}\phi_e + {}^{Zn}\Delta^{Zn^{2+}}\phi_e + {}^{H^+}\Delta^{H_2,Pt}\phi_e + {}^{Pt,H_2}\Delta^{Pt}\phi_e$ (5)

salt bridge

ie. ~ liquid junction potential between $Zn^{2+}(aq)$|$H^+(aq)$

Given the definition of a potential difference across an interface in equation 2,

$$^{H^+}\Delta^{H_2,Pt}\phi_e = \phi_e(H^+, aq) - \phi_e(H_2, Pt)$$
$$= -\{\phi_e(H_2, Pt) - \phi_e(H^+, aq)\}$$
$$= -\,^{Pt,H_2}\Delta^{H^+}\phi_e$$

and similarly,

$$^{Pt,H_2}\Delta^{Pt}\phi_e = -\,^{Pt}\Delta^{Pt,H_2}\phi_e$$

So equation 5 can equally well be written,

$$E = \{^{Pt}\Delta^{Zn}\phi_e + \,^{Zn}\Delta^{Zn^{2+}}\phi_e\} - \{^{Pt,H_2}\Delta^{H^+}\phi_e + \,^{Pt}\Delta^{Pt,H_2}\phi_e\} \quad (6)$$
$$= E^{\ominus}(\text{cell})$$

if the cell is set up under *standard* conditions (as indicated in Figure 5).

- Now write an expression for $E^{\ominus}$(cell) in terms of the standard electrode potentials of the two couples involved.
- From Block 7,

  $$E^{\ominus}(\text{cell}) = E^{\ominus}_{\text{RHE}} - E^{\ominus}_{\text{LHE}}$$
  $$= E^{\ominus}(Zn^{2+}|Zn) - E^{\ominus}(\text{S.H.E.}) \quad (7)$$

  where S.H.E. refers to the standard hydrogen electrode.

But *by convention*, $E^{\ominus}$(S.H.E.) = 0. Comparing the expressions in equations 6 and 7 suggests that the force of this convention is to set the potential difference terms involving the standard hydrogen electrode (those in the second curly bracket in equation 6) equal to zero; that is,

$$^{Pt,H_2}\Delta^{H^+}\phi_e + \,^{Pt}\Delta^{Pt,H_2}\phi_e = 0 \text{ (for the S.H.E., by convention)}$$

This, in turn, implies that the standard electrode potential of the (Zn^{2+}|Zn) electrode, *measured relative to the S.H.E. (and using platinum connecting leads)* is given by the following expression,

$$E^{\ominus}(Zn^{2+}|Zn) = \,^{Pt}\Delta^{Zn}\phi_e + \,^{Zn}\Delta^{Zn^{2+}}\phi_e$$

or more generally (that is, if $a(Zn^{2+}) \neq 1$):

$$E(Zn^{2+}|Zn) = \,^{Pt}\Delta^{Zn}\phi_e + \,^{Zn}\Delta^{Zn^{2+}}\phi_e \quad (8)$$

In other words, the electrode potential is the sum of *two* potential differences – the equilibrium potential difference across the metal–solution interface plus the equilibrium potential difference across a platinum–metal interface.

Thus, even if we know the electrode potential of a particular half-cell, we still can't determine the corresponding value of the equilibrium potential difference across the metal–solution interface. This reinforces our earlier conclusion:

Values for potential differences across single interfaces cannot be determined experimentally.

SAQ 1 Derive an expression for the potential difference across the following electrochemical cell considering all the interfaces involved:

Pt(s)|Cu(s)|Cu^{2+}(aq)|Zn^{2+}(aq)|Zn(s)|Pt(s)

Hence derive an expression for the emf of the cell in terms of the electrode potentials of the two half-cells.

2.3 Variation of potential with distance

You should now be convinced that a potential difference can develop at *any* phase boundary; if the phase boundary is between a metal and its aqueous solution, then the potential difference can be substantial (typical values are calculated to be of the order of 1 V). Let's now take a closer look at the double layer structure of the interface.

One of the earliest models of the electrified interface was proposed by H. Helmholtz in 1879. He suggested that solvated positive ions become 'lined up' in the solution at a *fixed* distance from the electrode. Each excess positive ion was imagined to be 'paired' with an excess electron at the surface of the metal. The effect is illustrated schematically in Figure 6. According to the Helmholtz model, the row nearest to the metal is composed mainly of water molecules, their electric dipoles (represented by arrows in Figure 6) orientated by the electric field. The next row is composed mainly of solvated positive ions: the line passing through the centres of the solvated ions is called the **outer Helmholtz plane** (**OHP**).* In this simplest of models the excess positive charge at the OHP is taken to be equal and opposite to the excess negative charge on the surface of the metal. Thus, the electric field arising from this charge separation is confined to the region between the metal surface and the OHP. A sketch of potential versus distance, for this particular model, is shown in Figure 7. This figure shows that the potential increases linearly with distance from the electrode surface up to the OHP; thereafter the potential remains constant. Thus, according to this model, the potential difference between the ions at the OHP and the bulk of the solution is zero.

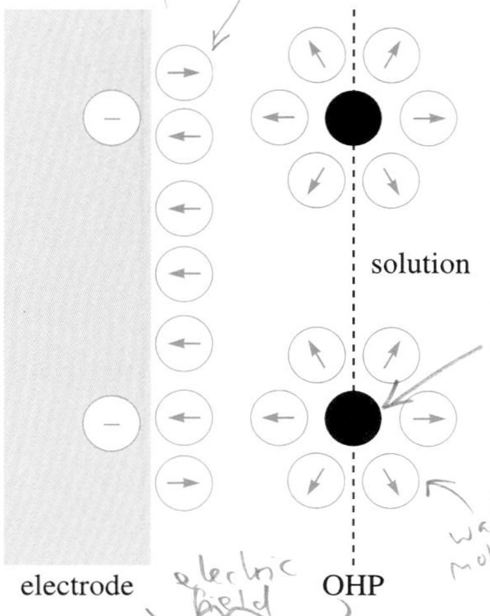

Figure 6 The Helmholtz model of the electrified interface between a metal and a solution.

The OHP lies at a distance of approximately one nanometre (10^{-9} m) from the metal surface; in other words, it is very close indeed to the surface! Over this very small distance the electrical potential changes typically by 1 V, which must mean that ions near electrodes are subjected to great forces.

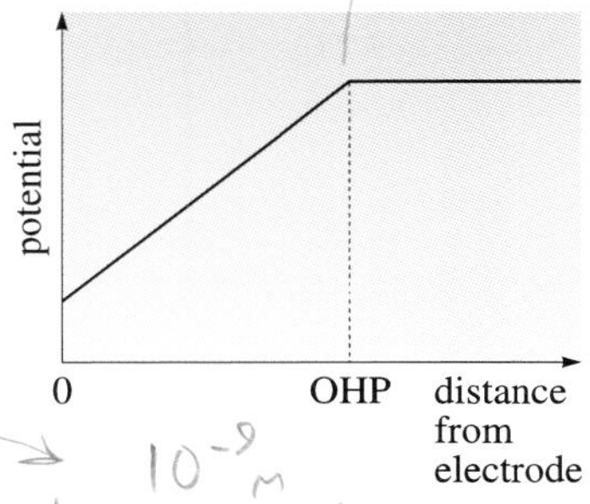

Figure 7 The variation in potential with distance from the electrode for the Helmholtz model. In line with the definition of electrical potential given in Block 7, the excess negative charge on the surface of the electrode means that it has a *lower* potential than the OHP.

As you might imagine, the Helmholtz picture outlined above is a very simplified one; even at an intuitive level it seems unlikely that the plot of potential versus distance should change so abruptly at the OHP. Indeed, later models have modified this treatment. However, we shall find it sufficient for our purposes.

2.4 Summary of Section 2

1 Potential differences arise at all interfaces. When the interface is between a metal and a solution containing its aqueous ions, the potential difference can arise through charge transfer reactions.

2 A change in the potential difference across a metal–solution interface can have a dramatic effect on the rate of an electrochemical reaction.

3 Any measurable potential difference involves more than one interface. The potential difference across a single interface cannot be measured experimentally. The equilibrium potential difference, $\Delta\phi_e$, across a metal–solution interface does not have the same value as the electrode potential.

4 The potential difference across a metal–solution interface is confined to a region very close to the electrode surface. In the Helmholtz model the potential changes linearly between the electrode surface and the outer Helmholtz plane.

* The term suggests that there is such a thing as an inner Helmholtz plane, and indeed there is, but we need not consider it.

3 THE EFFECT OF THE POTENTIAL DIFFERENCE ON THE RATE OF REACTION

3.1 Experimental behaviour

The rate of an electrochemical reaction is profoundly affected by the potential difference operating across the metal–solution interface. To reveal this dependence of rate on potential difference experimentally for the single, isolated electrode that we are considering calls for some ingenuity! But, it can be done, as we shall see in Section 4. In such an experiment, the rate is measured directly as a **current density**, i, which is simply the current, I, divided by the area, A, of the electrode; that is $i = I/A$. In other words, the current density is a measure of the rate of charge transfer (recall that the SI unit of current, the amp $\text{A} = \text{C s}^{-1}$) *per unit area*, which reflects the fact that electrode processes are, by definition, *surface* reactions.

The potential difference across a single interface cannot be measured, as we have already discussed, but what can be measured is the *change* in the potential difference across an interface. This change in potential difference is expressed by a new term, called the *overpotential*, which is given the symbol η (Greek 'eta'). We shall define this term more rigorously later on. However, for now we ask you to accept that an experimental plot of i versus η will reveal the nature of the dependence of reaction rate on the potential difference across an interface. A typical experimental plot of i versus η for the reduction of metal ions, M^{n+}, to metal, M, is shown in Figure 8. (The reasons why this plot involves negative values of both i and η will become apparent in Section 3.2.) It is clear from this plot that the relationship between i and η is fairly complex. To simplify the analysis of Figure 8 we can restrict the discussion to η values less negative than x. An experimental plot of i versus η for the reduction of Cu^{2+} ions to Cu metal at 298.15 K, at small values of η, is shown in Figure 9. It was J. Tafel who, in 1905, revealed that the relationship shown here is of the form

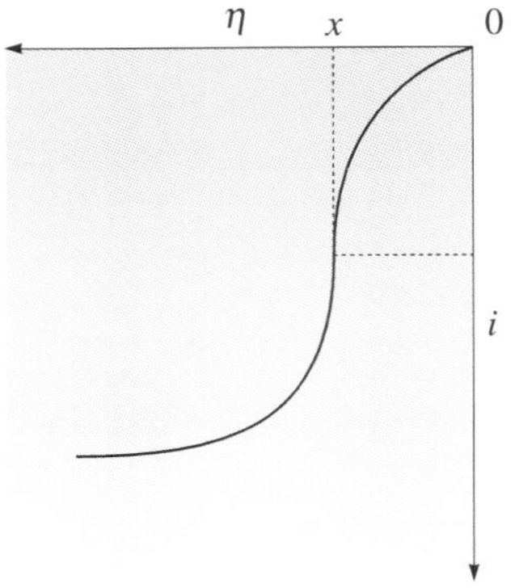

Figure 8 A typical experimental plot of i versus η for reduction of $M^{n+}(aq)$ to metal M(s).

$$\eta = a + b \log i \tag{9}$$

where a and b are constants for a given electrode system. Over this limited range of η, plots of η versus $\log i$, so called **Tafel plots**, are linear for many electrochemical systems: although the constant a changes from one system to another, values of b, the gradient of the plots, often bear a simple relationship to one another.

Our task now is to try to rationalize this Tafel-type behaviour. The best way to start is to look at a simple qualitative approach. This will allow us to introduce some of the sign conventions and will help explain terms like the overpotential.

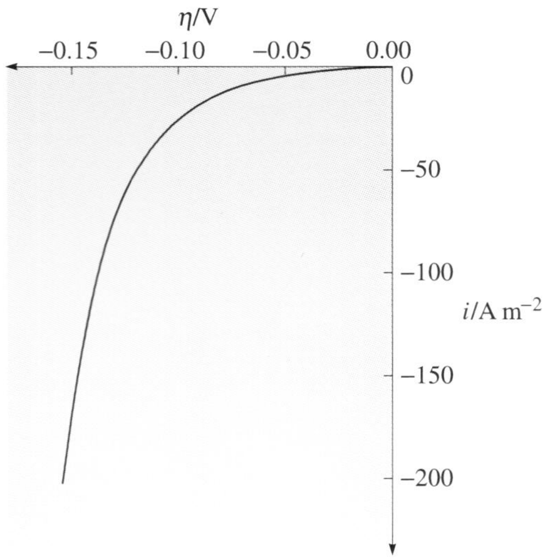

remember i is a direct measure of rate

Figure 9 An experimental plot of i versus η for reduction of $Cu^{2+}(aq)$ to Cu(s) at 298.15 K.

3.2 A simple qualitative approach to reaction rates

Consider once again the setting up of the zinc half-cell shown in Figure 1. On immersion of the zinc metal into the zinc solution, assume that the most likely reaction is oxidation of zinc metal (reaction 1 in the forward direction):

$$Zn(s) = Zn^{2+}(aq) + 2e \qquad (1)$$

STUDY COMMENT It is important that you now attempt Exercise 1 because it will give you a 'feel' for some important aspects of electrode kinetics. It also introduces two important definitions that will be used throughout the remainder of the Block. You are given some help with the first sketch you are asked to produce and, if you have any problems, a full commentary is provided in the answer to the Exercise.

EXERCISE 1 In this Exercise you are asked to sketch various graphs of one variable plotted against another, from the instant of immersion of the zinc metal until equilibrium is reached. Although you won't be able to show the exact nature of the curve, draw the one you think to be most likely. For the purposes of this Exercise, assume that the potential difference arises solely from charge transfer reactions. Label the axes of your graphs and any appropriate points. The sketches required are plots of:

(a) the rate of reaction 1 (an oxidation reaction) versus time (to help show you the sort of sketch required, this first one is shown in Figure 10);

(b) the rate of the reverse of reaction 1 (a reduction reaction) versus time;

(c) the *net* rate of the oxidation reaction versus time, where this is defined as follows:

$$\text{net rate of oxidation reaction} = (\text{rate of oxidation reaction}) - (\text{rate of reduction reaction}) \qquad (10)$$

(d) the variation of the potential difference versus time, where the potential difference, $\Delta\phi$, across a metal–solution interface is defined as follows:

$$\Delta\phi = {}^{M}\Delta^{S}\phi = \phi_M - \phi_S \qquad (11)$$

where ϕ_M is the potential of the metal and ϕ_S is the potential of the solution.

Be careful with signs. In this case, the potential difference, as defined above, becomes more negative with time, as excess negative charge builds up on the electrode and excess positive charge builds up in the solution alongside it.

(e) the variation of the *net* oxidation rate versus potential difference.

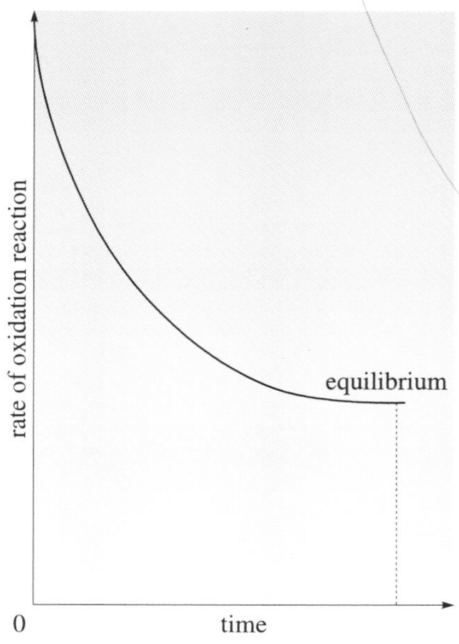

Figure 10 A sketch of the rate of the oxidation reaction versus time.

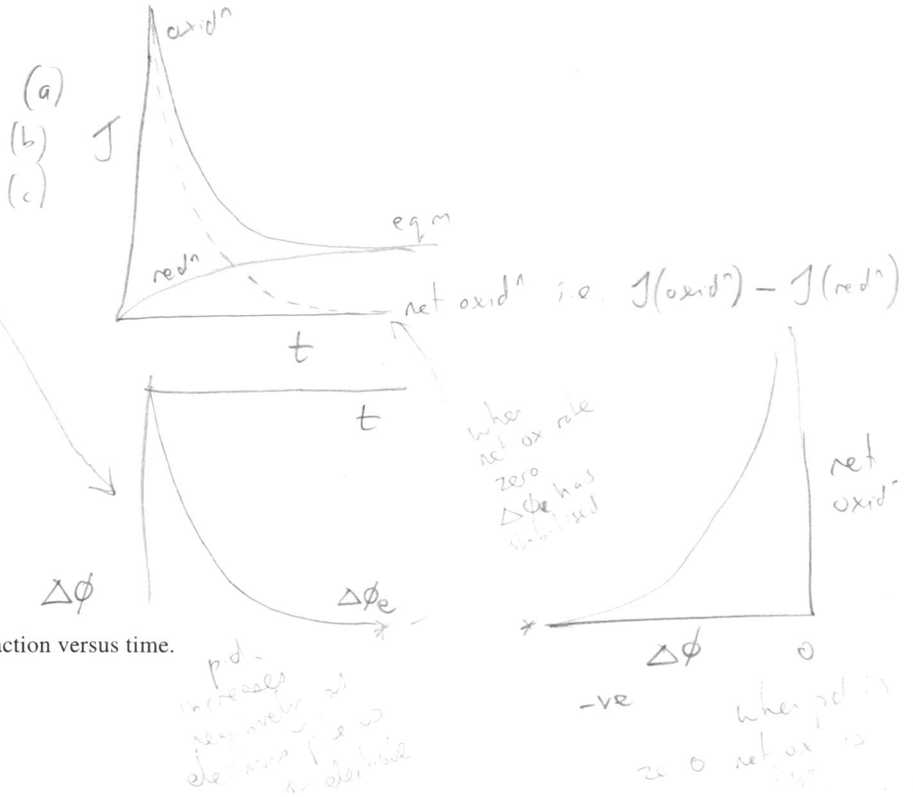

STUDY COMMENT Now compare your sketches with the ones shown in the answer to Exercise 1 (Figures 45–49, on p. 65).

The sketch shown in Figure 49 (repeated here as Figure 11) has been produced over a very limited potential difference range: we have followed the process as a zinc electrode is immersed in an aqueous solution containing zinc ions and the potential difference has changed from zero to $\Delta\phi_e$. The process is spontaneous and no external circuit is required. However, we could connect the zinc electrode (if we were to make it part of an electrochemical cell) to an external circuit and so, for example, push more electrons onto the zinc metal.

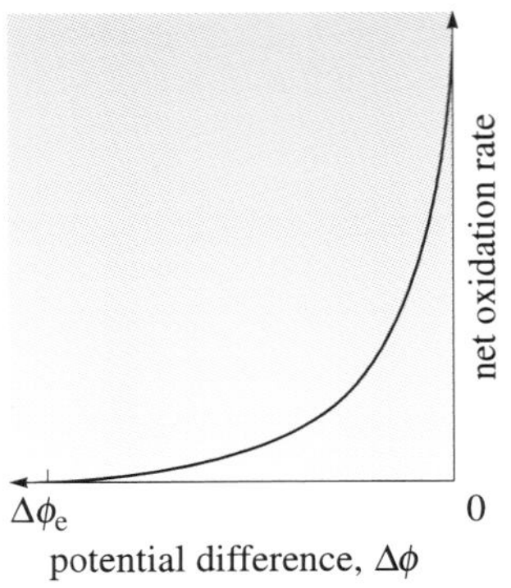

Figure 11 A sketch of the variation of *net* oxidation rate versus the potential difference across the metal–solution interface ($\Delta\phi$) for the setting up of a zinc half-cell. Here, $\Delta\phi_e$ is the equilibrium potential difference.

- What will this do to the value of $\Delta\phi$?
- When the system is at equilibrium, the metal has a negative charge and the solution a positive charge. The value of $\Delta\phi$, given by

  $$\Delta\phi = \phi_M - \phi_S$$

  will be negative. Pushing more electrons onto the metal will make ϕ_M even more negative than ϕ_S, thereby making the potential difference more negative.

- Which process will now dominate, oxidation or reduction?
- As the metal is made more negative relative to the solution the rate of the reduction reaction will increase, so, as the potential difference becomes more negative than $\Delta\phi_e$, the rate of the reduction reaction will exceed the rate of the oxidation reaction. *Therefore, the net oxidation rate (as defined in equation 10) will become negative.*

Thus Figure 11 can now be extended to include negative values of the net oxidation rate, as shown in Figure 12.

In a similar manner, the external circuit could be used to withdraw electrons from the zinc electrode. Starting from the equilibrium position this will make the potential difference *less negative* and the rate of the oxidation reaction will exceed the rate of the reduction reaction such that the net oxidation rate will become more positive. As more and more electrons are withdrawn from the zinc metal, the potential difference will eventually reach zero (as it was at the moment of immersion of the zinc metal in the solution, before any external circuit was connected). As even more electrons are withdrawn, the potential difference will become positive. A graph of net oxidation rate versus potential difference over this extended range is sketched in Figure 13.

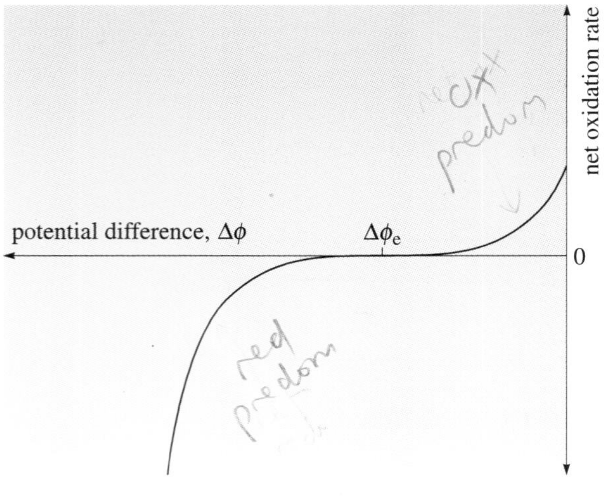

Figure 12 A sketch of the variation of *net* oxidation rate versus potential difference over an extended range of negative values of the potential difference, $\Delta\phi$.

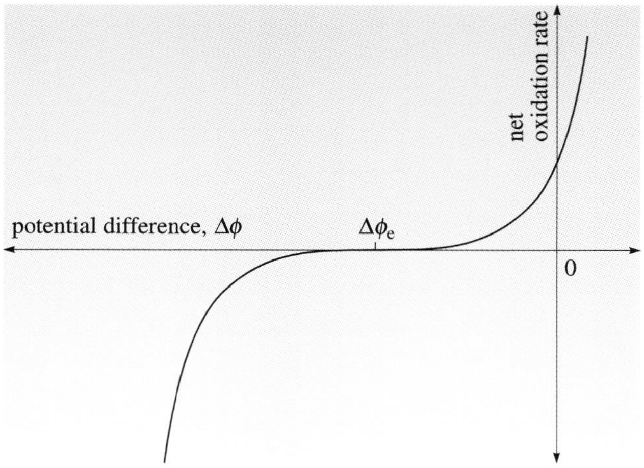

Figure 13 A sketch of the variation of net oxidation rate versus potential difference over an extended range that includes both negative and positive values of $\Delta\phi$.

Part of Figure 13 now looks like the experimental plot shown in Figure 9, but the axes are labelled differently. However, we can change this. In an experimental context, it is the *net* rate of the electrode reaction that is measured directly as the current density (i), so the vertical axis in Figure 13 can certainly be labelled i.

Of course, whether the current density is positive or negative depends on the convention we choose to adopt. We shall adopt the following convention:

When there is net oxidation the current density is taken to be positive.

So the vertical axis of Figure 13 can be labelled i, and has the correct orientation with positive values of i above the horizontal axis.

At this point it might aid understanding if we introduce a notation that allows us to distinguish between the current densities of the oxidation and reduction reactions taking place at the electrode. If we represent the current density associated with the oxidation reaction as i_{ox} and the corresponding current density associated with the reduction reaction as i_{red}, then, given the above convention, the *measured* current density, i, is given by

$$i = i_{net} = i_{ox} - i_{red} \quad (12)$$

Clearly, i is positive if $i_{ox} > i_{red}$, implying that net oxidation is taking place at the electrode. Conversely, if $i_{red} > i_{ox}$, net reduction is taking place at the electrode and the measured current density is then assigned a negative value.

What about the horizontal axis in Figure 13? This is labelled 'potential difference' (implicitly across the metal–solution interface), but we have said that values of this quantity cannot be determined experimentally. As the experimental plot in Figure 9 suggests, the way around this problem lies with the **overpotential**, η ; formally, it is defined as follows:

$$\eta = \Delta\phi - \Delta\phi_e \quad (13)$$

where $\Delta\phi$ is the potential difference across the interface and $\Delta\phi_e$ is the corresponding *equilibrium* potential difference across the interface.

As we shall see in Section 4, this *difference* in potential differences *can* be measured experimentally.

- Given the definition of the overpotential in equation 13, try to convert Figure 13 into a sketch of current density (i) against overpotential (η).

■ Conversion of the vertical axis is straightforward, as stated above. All we need to do is replace 'net oxidation rate' with 'i'. For the horizontal axis, consider the following: equation 13 shows that the origin of the required plot, the point for which $\eta = 0$, corresponds to the point $\Delta\phi = \Delta\phi_e$ on the horizontal axis in Figure 13. The direction of the η axis can be established by noting that when $\Delta\phi = 0$ then $\eta = -\Delta\phi_e$ (equation 13). This value of η will be *positive* because $\Delta\phi_e$ is negative (Figure 13). Therefore a sketch of current density versus overpotential should resemble Figure 14. Essentially all that's happened is that we've shifted the curve *along* the horizontal axis.

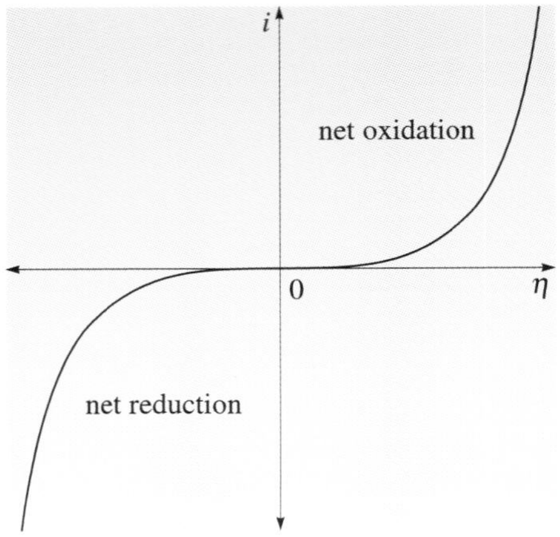

Figure 14 A sketch of current density versus overpotential for both positive and negative values of the overpotential.

Now you can see the link between Figure 9 and Figure 14. Figure 9 is an experimental plot for *net reduction only*; this forms the lower left-hand quadrant of Figure 14. An experimental plot for net oxidation should resemble the upper, right-hand quadrant of Figure 13. Note that the sign of the overpotential dictates the net process:

becuse the ϕ is going down, ∴ receiving e⁻

Negative overpotentials lead to net reduction, and positive overpotentials lead to net oxidation.

ϕ inc therefore losing e⁻

Notice the bold letters: this might help you to remember the connection.

This simple (although you may dispute this!) qualitative account of electrochemical reaction rates has taken us some considerable way towards an understanding of the variation of rate with potential difference, introduced us to some of the terminology used and shown us how such data are presented. However, although we have the axes labelled correctly, we have only guessed at the *shape* of the curve of rate versus potential difference.

10/9/09
11.30am
stop
here

SAQ 2 State whether the following statements concerning a standard hydrogen electrode are true or false.

(a) At the instant of immersion of a platinum electrode into a standard solution of hydrogen ions at 298.15 K (with hydrogen gas present at 1 bar), the overpotential is 0.0 V.

(b) Under equilibrium conditions at 298.15 K, the interfacial potential difference is 0.0 V.

(c) Under equilibrium conditions at 298.15 K, the overpotential is 0.0 V.

(d) When the overpotential is negative, the only reaction taking place will be:

$$H^+(aq) + e = \tfrac{1}{2}H_2(g)$$

(e) When the overpotential is 0.0 V, the rate of the reduction reaction and the rate of the reverse, oxidation, reaction are both zero.

have highest value here.
T X whoops.
F ✓
T ✓
f ✓
F ✓

3.3 The Butler–Volmer equation

The exact nature of the curve shown in Figure 14 is predicted by the **Butler–Volmer equation**. The equation has a sound theoretical basis and can be developed by considering how a change in potential difference affects the activation energy for the charge transfer processes (both oxidation and reduction) at a single electrode. However, for our purposes, it is sufficient to state the equation. In its general form, the equation reads as follows:

$$i = i_e\left\{\exp\left(\frac{\alpha_{ox}F\eta}{RT}\right) - \exp\left(\frac{-\alpha_{red}F\eta}{RT}\right)\right\} \qquad (14)$$

where F is the Faraday constant (96 485 C mol^{-1}), α_{ox} and α_{red} are called *transfer coefficients*, and i_e is known as the *exchange current density*.

Some insight into the meaning of the various quantities in equation 14 can be obtained by rewriting it as follows:

$$i = \left\{i_e\exp\left(\frac{\alpha_{ox}F\eta}{RT}\right)\right\} - \left\{i_e\exp\left(\frac{-\alpha_{red}F\eta}{RT}\right)\right\} \qquad (15)$$

and recalling our definition of the net current density,

$$i = i_{ox} - i_{red} \qquad (12)$$

It then becomes apparent that the terms in curly brackets in equation 15 are expressions for the oxidation and reduction current densities at the electrode; that is,

$$i_{ox} = i_e\exp\left(\frac{\alpha_{ox}F\eta}{RT}\right) \qquad (16)$$

and

$$i_{red} = i_e\exp\left(\frac{-\alpha_{red}F\eta}{RT}\right) \qquad (17)$$

- What happens to the expressions in equations 16 and 17 *at equilibrium*; that is, when $\eta = 0$?
- Since $e^0 = 1$ and $e^{-0} = 1$ (try this on your calculator), equations 16 and 17 together imply that

$$i_{ox} = i_{red} = i_e \textit{ at equilibrium} \qquad (18)$$

which is commonsense

In other words, i_e is a measure of the rate of the oxidation (or the reduction) reaction at the electrode at equilibrium: it is called the **exchange current density**.

The quantities represented by α in equations 14–17 are called **transfer coefficients**: α_{ox} is the transfer coefficient for the oxidation reaction at the electrode, and α_{red} is the transfer coefficient for the corresponding reduction reaction. Typically, values of α are 0.5, 1.0, 1.5 or 2.0. As you will see later on (Section 5.2), these values provide insight into the possible mechanisms of the oxidation and reduction reactions at the electrode.

A plot of i versus η derived from the Butler–Volmer equation does indeed have the form of the curve shown in Figure 14. Check this for yourself if you wish: calculate i for the situation in which, say, i_e is 10^{-8} A m^{-2}, $\alpha_{ox} = \alpha_{red} = 0.5$ and η varies from 0.3 V through 0 to – 0.3 V in steps of 0.1 V. Figure 14 has been produced for the situation in which $\alpha_{ox} = \alpha_{red}$: if α_{ox} has a different value from α_{red} then the curve becomes less symmetrical than the one shown.

More usefully, however, the Butler–Volmer equation can be used to obtain the important quantities, i_e, α_{ox} and α_{red} from experimental Tafel plots in the following way.

Consider the full Butler–Volmer equation (equation 14 or 15). At large, positive values of the overpotential (but still less than the magnitude of x in Figure 8) the *first* exponential term in the equation becomes very large relative to the second. For example, if $\eta = +0.3$ V at 300 K with $\alpha_{ox} = \alpha_{red} = 0.5$, then the first term becomes approximately e^6 (≈ 400) and the second term e^{-6} ($\approx 2.5 \times 10^{-3}$). You will have discovered this for yourself if you tried the calculation mentioned above.

Under these conditions it is a reasonable approximation to neglect the second term, so equation 14 (or 15) becomes

$$i = i_e \exp\left(\frac{\alpha_{ox} F \eta}{RT}\right) \text{ for large, positive values of } \eta \qquad (19)$$

which implies that $i \approx i_{ox}$ (see equation 16).

Thus, at large positive values of the overpotential the Butler–Volmer equation predicts that a plot of current density versus overpotential should yield an exponential curve. Under these conditions the net current density, i, will be positive (consistent with our sign convention), and the rate of the reduction reaction will be very small compared with the rate of the oxidation reaction.

Equation 19 can be related to Tafel plots by taking logarithms

$$\ln i = \ln i_e + \frac{\alpha_{ox} F \eta}{RT}$$

$$\text{or } 2.303 \log i = 2.303 \log i_e + \frac{\alpha_{ox} F \eta}{RT}$$

Dividing though by 2.303 and rearranging then gives:

$$\log i = \frac{\alpha_{ox} F \eta}{2.303 RT} + \log i_e \qquad (20)$$

Since this equation has a linear form ($y = mx + c$), a plot of log i against η should be a straight line. Such a plot will yield a value for α_{ox} from the slope and a value of log i_e (and hence i_e) from the intercept on the log i axis (when $\eta = 0$). A comparable analysis for large negative values of the overpotential would yield values for i_e and α_{red}. However, Tafel plots tend to be presented with the overpotential on the *vertical* axis, and we shall adopt this form of presentation. Over both positive and negative values of the overpotential, experimental Tafel plots should look something like Figure 15; however, the two halves of the Tafel plots are often determined and discussed separately. Notice that the horizontal axis in Figure 15 is labelled as log $|i|$ rather than log i. This allows us to cope with negative values of the current density.

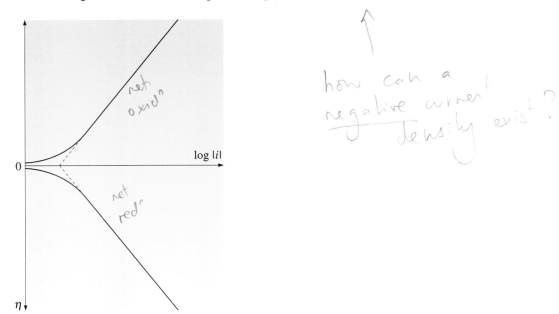

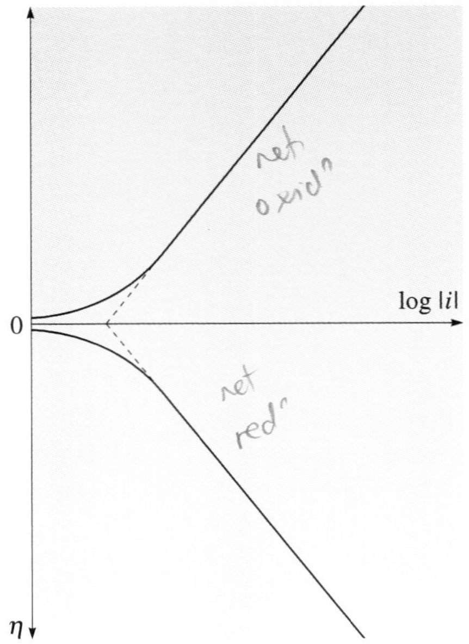

Figure 15 Schematic Tafel plots for both positive and negative values of the overpotential.

Notice also that, at small values of the overpotential, the plot is no longer linear: this is to be expected because when η is small the two exponential terms in equation 14 are of similar magnitude, and so the approximation used to derive equation 19 is no longer valid. As discussed above, the value of i_e is determined from the intercept of the linear portion of the plot with the horizontal axis – that is, with the $\log|i|$ axis. As shown in Figure 15, the values of i_e obtained in this way from the two Tafel plots must be the same for a particular electrode system under identical conditions. However, as already indicated, the value for α_{red} is not necessarily the same as α_{ox} for a particular electrode system. Thus, the magnitudes of the slopes of the two lines shown in Figure 15 could (and often do) differ.

■ How would you determine a value of α_{ox}, say, from the Tafel plots shown in Figure 15?

▪ α_{ox} would be determined from the slope of the linear portion of the upper plot shown in Figure 15, because net oxidation occurs when $\eta > 0$. But take care! The slope of this line is the *inverse* of the slope of the line represented by equation 20, because Figure 15 is a plot of η against $\log|i|$, *not* of $\log|i|$ against η.

This point is seen more clearly if we rearrange equation 20 by isolating the term that involves η on the left-hand side, as

$$\frac{\alpha_{ox}F\eta}{2.303\,RT} = \log i - \log i_e$$

so

$$\eta = \frac{2.303\,RT}{\alpha_{ox}F}\log i - \frac{2.303\,RT}{\alpha_{ox}F}\log i_e \qquad (21)$$

Thus the slope of the linear portion of the Tafel plot is given by

$$\text{slope} = \frac{2.303\,RT}{\alpha_{ox}F} \text{ for net oxidation } (\eta > 0) \qquad (22)$$

A comparable analysis for negative values of the overpotential leads to the following expression for the slope of the linear portion of the lower plot in Figure 15:

$$\text{slope} = -\frac{2.303\,RT}{\alpha_{red}F} \text{ for net reduction } (\eta < 0) \qquad (23)$$

where the negative sign can be traced back to the form of the second exponential term in the Butler–Volmer equation (equation 14 or 15).

In Section 6 (and for the remainder of the Course) we make extensive use of Tafel plots and of the i_e and α values determined from them. But we still haven't discussed how such data are determined. This is covered in the next Section.

SAQ 3 The following data were obtained for a platinum electrode of surface area 2 cm^2 in contact with $Fe^{2+}(aq)$ and $Fe^{3+}(aq)$ ions at $25\,°C$.

η/mV	+50	+100	+150	+200	+250
I/mA	8.8	25.0	58.0	131	298
i	44	125	290	655	1490

(a) Identify the electrode process.

(b) Construct a Tafel plot and determine the values of the transfer coefficient (α) and the exchange current density (i_e) for the electrode process. (Note that you are given the value of the *current*.)

2·82   3·17

logs

1·64   2·46

2·10

POSTER

$i = \dfrac{I}{A}$

$I = 8.8 \times 10^{-3}\ A$

$A = 2\ cm^2$

$i = \dfrac{8.8 \times 10^{-3}\ A}{2\ cm^2}$

$i = 4.4 \times 10^{-3}\ A\ cm^{-2}$

$= 44\ A\ m^{-2}$

$(1000\ cm^2 = 1\ m^2)$

3.4 Summary of Section 3

By focusing on the events at a single metal–solution interface, we have now introduced most of the terminology, sign conventions, and important general results that we shall need in discussing the kinetics of electrode processes. The key points are collected in Box 1 and Figure 16.

Box 1

1 For electrode processes, the observed dependence of reaction rate on the potential difference, $\Delta\phi$, across the electrode–solution interface can be recorded as a plot (see Figure 16) of *net* current density, i,

$$i = I/A$$

where I is the net current and A is the area of the electrode, against overpotential, η:

$$\eta = \Delta\phi - \Delta\phi_e \qquad (13)$$

where $\Delta\phi_e$ is the potential difference at equilibrium.

2 By convention:

$$\Delta\phi = \phi_M - \phi_S \qquad (11)$$

and

$$i = i_{ox} - i_{red} \qquad (12)$$

where i_{ox} and i_{red} are the current densities for the oxidation and reduction reactions, respectively, at the electrode.

3 The *net* process at a single electrode is dictated by the *sign* of the overpotential, as shown in Figure 16. *At equilibrium*, both the overpotential and the net current density are zero ($\eta = 0$ and $i = 0$); then $i_{ox} = i_{red} = i_e$, the exchange current density.

4 The general shape of the plot in Figure 16 is consistent with the predictions of the Butler–Volmer equation:

$$i = i_e \left\{ \exp\left(\frac{\alpha_{ox} F\eta}{RT}\right) - \exp\left(\frac{-\alpha_{red} F\eta}{RT}\right) \right\} \qquad (14)$$

where α_{ox} and α_{red} are the transfer coefficients for the oxidation and reduction reactions, respectively, at the electrode.

5 Values of the quantities i_e, α_{ox} and α_{red} can be determined from experimental Tafel plots (see Figure 15) – plots of overpotential (on the vertical axis) against $\log|i|$. A Tafel plot is usually a straight line, except at very small values of η:

- Projecting the linear portion of an experimental Tafel plot onto the $\log|i|$ axis yields the value of i_e.
- The gradient of the linear portion of an experimental Tafel plot yields the value of α_{ox} or α_{red}, where

$$\text{slope} = \frac{2.303\,RT}{\alpha_{ox} F} \quad \text{or} \quad -\frac{2.303\,RT}{\alpha_{red} F}$$

net oxidation ($\eta > 0$) | net reduction ($\eta < 0$)

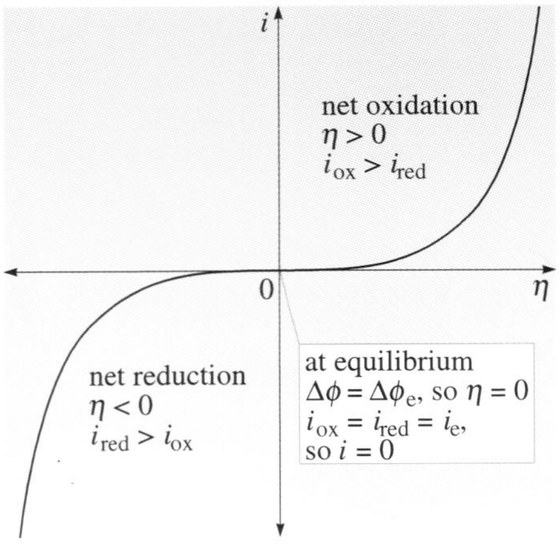

Figure 16 A sketch of net current density versus overpotential for a typical electrode process, incorporating some of the terminology and conventions introduced in Section 3.

4 OBTAINING EXPERIMENTAL DATA

The dependence of current density on the overpotential is one of the central themes of electrode kinetics. The dependence is expressed theoretically by the Butler–Volmer equation and empirically, under particular conditions, by the Tafel relationship. But how do we determine experimental values of overpotential and current density for a single electrode? As mentioned earlier we cannot, of course, measure quantities such as the current and the potential difference of a single *isolated* electrode; for one thing, current doesn't *flow* in an isolated electrode. The quantities are measured experimentally by making the single electrode part of an electrochemical cell. But if the conditions are kept constant at the other electrode, changes in current and potential difference *can* be related to the processes occurring at the electrode of interest. However, to operate this system, it is often necessary to add a third electrode.

To see why this should be so, think back to how we measured a value for the electrode potential of a particular system. For zinc metal in contact with aqueous zinc ions, for example, we could set up the system shown in Figure 5, but replace the DVM by an ammeter and a variable potential source, as shown in Figure 17a.

When a potentiometer circuit is adjusted so that no current flows through the cell and external circuit, the recorded potential difference across the cell, V_1, can be equated with the electrode potential of the $Zn^{2+}|Zn$ half-cell: in symbols (from Section 2.2):

$$V_1 = E = {}^{\mathrm{Pt}}\Delta^{\mathrm{Zn}}\phi_{\mathrm{e}} + {}^{\mathrm{Zn}}\Delta^{\mathrm{Zn^{2+}}}\phi_{\mathrm{e}} \qquad (24)$$

A potential difference across the zinc–solution interface that is different from the equilibrium potential difference will result in net oxidation or net reduction at the interface, with a consequent flow of current in the external circuit. To impose this new potential difference and measure the resulting current we retain the system in Figure 17a, but add a further electrode, as shown in Figure 17b.

Imposing a potential difference across the two right-hand electrodes changes the potential difference across the zinc–solution interface from its equilibrium value: this causes a current to flow in the right-hand circuit, which is recorded on ammeter 2. By rebalancing the potentiometer circuit (the left-hand circuit), so that no current flows according to ammeter 1, the new potential difference V_2 across the two left-hand electrodes can be recorded.

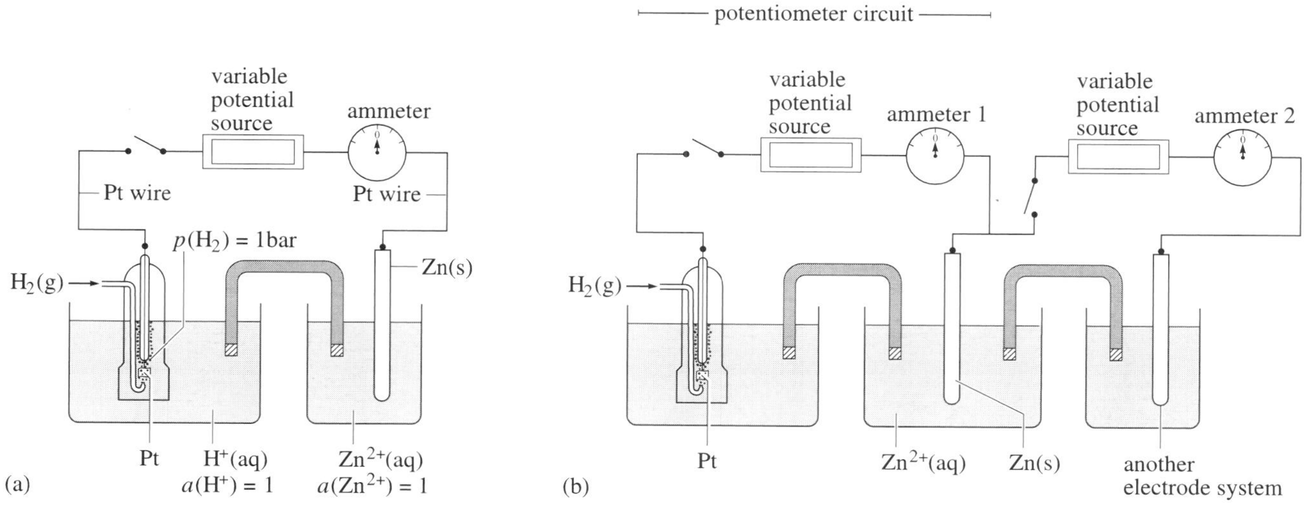

Figure 17 (a) Repeat of Figure 5 with the DVM replaced by an ammeter and a variable potential source; (b) the system required to measure the current at various values of the overpotential.

With no current flowing through the hydrogen electrode, the potential differences at this electrode will still be at their equilibrium values (and, by convention, can be set at zero), so:

$$V_2 = {}^{\mathrm{Pt}}\Delta^{\mathrm{Zn}}\phi + {}^{\mathrm{Zn}}\Delta^{\mathrm{Zn}^{2+}}\phi \qquad (25)$$

From equations 24 and 25,

$$V_2 - V_1 = {}^{\mathrm{Zn}}\Delta^{\mathrm{Zn}^{2+}}\phi - {}^{\mathrm{Zn}}\Delta^{\mathrm{Zn}^{2+}}\phi_{\mathrm{e}} + {}^{\mathrm{Pt}}\Delta^{\mathrm{Zn}}\phi - {}^{\mathrm{Pt}}\Delta^{\mathrm{Zn}}\phi_{\mathrm{e}} \qquad (26)$$

However, experimental evidence suggests that the potential difference across a metal-metal interface is virtually independent of the imposed, external potential difference and so $({}^{\mathrm{Pt}}\Delta^{\mathrm{Zn}}\phi - {}^{\mathrm{Pt}}\Delta^{\mathrm{Zn}}\phi_{\mathrm{e}})$ can be put equal to zero. Thus, equation 26 becomes

$$V_2 - V_1 = {}^{\mathrm{Zn}}\Delta^{\mathrm{Zn}^{2+}}\phi - {}^{\mathrm{Zn}}\Delta^{\mathrm{Zn}^{2+}}\phi_{\mathrm{e}} = \eta \qquad (27)$$

In other words, the overpotential is determined by noting the two readings of potential difference, V_1 and V_2. The current is determined directly from the readings on ammeter 2. The current *density* is determined by dividing the current by the surface area of the zinc electrode. In principle, the surface area of the electrode can be obtained by the procedures outlined in Block 5. When the electrode is a polished metal sheet, however, the area is often determined simply from measurement of its length and breadth (although this does seem a little crude!).

There are problems associated with these measurements (especially in dilute solutions), but often these can be overcome with careful design of the equipment and appropriate choice of conditions. For example, as soon as current starts to flow in a system, a potential difference arises purely because of the resistance of the solution. The resistance of the solution is minimized by adding large amounts of inert but highly conducting substances such as potassium chloride. Ions from these salts effectively transport the current through the cell, but do not take part in the electrode reactions because under most conditions they are not easily oxidized or reduced. This process of adding inert substances is a favourite trick of electrochemists.

Naturally, in laboratories where such data are collected routinely, the measurements are somewhat more automated than is implied above. For example, in the 'potentiodynamic method' the potential of the electrode of interest is steadily changed such that the overpotential varies smoothly from a high negative value, through zero to a high positive value, the current being recorded at various values of the overpotential. Remember that, according to our convention, the current is awarded a positive sign when the overpotential is positive (implying net oxidation); when the overpotential is negative (net reduction), the current is given a negative sign.

5 USING TAFEL DATA

As already described, Tafel plots yield values of the exchange current density, i_e, and the values of α (α_{ox}, α_{red} or both). Used independently, these values give us important information on electrode processes. When used together, they prove invaluable in understanding many electrochemical processes – as we shall see throughout the remainder of this Block and in Topic Study 3.

5.1 i_e values

Table 1 lists a sample of i_e values mainly determined from Tafel plots of experimental data obtained under standard concentration conditions at 298.15 K. Remember that i_e values provide a measure of the rate of the electrode reaction at equilibrium. Notice from Table 1 the enormous range of values of i_e even for one electrode reaction: for example, the hydrogen reduction reaction is 10^9 times faster on a platinum surface than on a mercury surface. Thus, if you wanted to encourage this particular electrode process you would tend to select platinum or palladium for your electrode material (if cost was no object!); if you wanted to discourage this particular electrode process, then you could select lead or mercury as your electrode material. To be specific, if you are trying to measure the electrode potential of a system relative to a standard hydrogen electrode you need to ensure that the reactions occurring at the hydrogen electrode are relatively fast so that a balance point, with no current flow, can be achieved quickly and easily: this is one reason why platinum is selected as the electrode material for the hydrogen electrode. On the other hand, if you are attempting to extract a metal from an *aqueous* solution of its ions, then the two possible competing processes at the cathode need to be considered – reduction of the metal ions and reduction of hydrogen ions. The hydrogen reduction reaction can be effectively quashed by using a mercury cathode with which many elements (but not hydrogen) form some sort of amalgam (solution in mercury as solvent). We shall discuss this further in Section 7.2.

Table 1 Approximate exchange current density values at 298.15 K and under standard concentration conditions (defined as c = 1 mol dm^{-3}).

Reaction	Electrode material	i_e/A m^{-2}	Reaction	Electrode material	i_e/A m^{-2}
$H^+(aq) + e = \frac{1}{2}H_2(g)$	Pd	10	$Ag^+(aq) + e = Ag(s)$	Ag	10^4
	Pt	10	$Fe^{3+}(aq) + e = Fe^{2+}(aq)$	Pt	10
	Rh	1		Rh	10
	Ir	1		Ir	10
	Ni	10^{-1}		Pd	10
	Au	10^{-1}	$Ce^{4+}(aq) + e = Ce^{3+}(aq)$	Pt	1
	Fe	10^{-2}	$Cr^{3+}(aq) + e = Cr^{2+}(aq)$	Hg	10^{-2}
	Cu	10^{-2}	$Ni^{2+}(aq) + 2e = Ni(s)$	Ni	10^{-5}
	Ag	10^{-2}	$Zn^{2+}(aq) + 2e = Zn(s)$	Zn	10^{-3}
	W	10^{-2}	$Cd^{2+}(aq) + 2e = Cd(s)$	Cd	10^3
	Mo	10^{-3}	$Cu^{2+}(aq) + 2e = Cu(s)$	Cu	10
	Nb	10^{-3}	$Fe^{2+}(aq) + 2e = Fe(s)$	Fe	10^{-4}
	Sn	10^{-4}	$O_2(g) + 4H^+(aq) + 4e = 2H_2O(l)$	Pt	10^{-6}
	Ti	10^{-4}		Fe	10^{-10}
	Zn	10^{-6}			
	Al	10^{-6}			
	Cd	10^{-7}			
	Mn	10^{-7}			
	Tl	10^{-7}			
	Pb	10^{-8}			
	Hg	10^{-8}			

This variation of i_e for a given electrode reaction with the nature of the electrode material constitutes the subject of **electrocatalysis**: the electrode is acting as a catalyst for the charge transfer reaction. Put at its simplest, the higher the value of i_e for a given reaction, the better is the corresponding electrode as an **electrocatalyst** for that reaction.

The values of i_e listed in Table 1 are for *standard* concentration conditions (defined in the present context as being $c = 1$ mol dm^{-3} rather than $a = 1$) at 298.15 K and, of course, different values would be expected under different conditions. For the $Ag^+(aq) + e = Ag(s)$ reaction, for example, the rate of the reaction, and hence i_e, is bound to be influenced by both the temperature and the concentration of the Ag^+ species in solution. Increasing the temperature or increasing the concentration of the silver ions should increase the rate of the reaction. In practice, however, the actual relationship between concentration and i_e is not straightforward and, as you might expect from your knowledge of solution chemistry, the value of i_e is sometimes affected by changes in the concentrations of species that do not appear in the stoichiometric equation. Information would be required on the mechanism of the reaction and the nature of the rate-limiting step. In addition, the measured values of i_e on a given electrode are frequently highly dependent on the pre-treatment and physical form of the electrode. As with purely 'chemical' heterogeneous catalysis, this again reflects the sensitivity of any surface reaction to the precise state, morphology and composition of the surface.

5.2 α values

Transfer coefficients, α values, can be determined from the slopes of the Tafel plots as discussed in Section 3.3: α_{ox} values are obtained from plots determined using positive values of the overpotential, and α_{red} values from plots using negative values. The relationship between the slope of the Tafel plot and the corresponding α values (determined using equations 22 or 23), for some typical Tafel slopes at 300 K are given in Table 2.

Table 2 Relationship between Tafel slope and α values at 300 K.

Tafel slope/mV	α_{red} or α_{ox}
± 120	0.5
± 60	1.0
± 40	1.5
± 30	2.0

The value of α is related to the *mechanism* of an electrochemical reaction. *For a reaction scheme written showing overall reduction*, it can be shown that the relationship is given by the equations:

$$\alpha_{red} = \gamma_B + 0.5n \qquad (28)$$

and

$$\alpha_{ox} = \gamma_A + 0.5n \qquad (29)$$

where γ_B = the number of steps involving electron transfer *before* the rate-limiting step; γ_A = the number of steps involving electron transfer *after* the rate-limiting step; and n = the number of electrons transferred *in* the rate-limiting step (0 or 1).

Let us consider a particular example. A reaction that has received a lot of attention from a mechanistic point of view is the following:

$$H^+(aq) + e = \tfrac{1}{2}H_2(g) \qquad (30)$$

Although many steps can be considered, one of the proposed mechanisms involves just two steps:

Mechanism I

$$H^+(OHP) + e \longrightarrow H(ad) \qquad (31)$$

$$H(ad) + H^+(OHP) + e \longrightarrow H_2(ad) \qquad (32)$$

In the first step (reaction 31) an electron from the electrode surface 'jumps' to the approaching hydrogen ion when the ion is somewhere between the OHP and the electrode surface; the resulting hydrogen atom is adsorbed on the electrode surface at an active centre. In the second step (equation 32) an adsorbed hydrogen atom comes into contact with a hydrogen atom being formed by the interaction of an electron with a hydrogen ion from the OHP.

A possible alternative mechanism is as follows:

Mechanism II

$$H^+(OHP) + e \longrightarrow H(ad) \qquad (31)$$

$$H^+(OHP) + e \longrightarrow H(ad) \qquad (31)$$

$$2H(ad) \longrightarrow H_2(ad) \qquad (33)$$

Here, the first step is the same as in Mechanism I, but two adsorbed hydrogen atoms then come together on the electrode surface to form an adsorbed hydrogen molecule (reaction 33). If this surface reaction is the rate-limiting step, the first step has to occur twice, which is why we have repeated reaction 31 in writing Mechanism II.

■ Use equation 28 to calculate values of α_{red} for the two proposed mechanisms considering the various possibilities (reactions 31, 32 or 33) for the rate-limiting step.

■ Equation 28 yields the values of α_{red} collected in Table 3.

Table 3 Theoretical values of α_{red} for Mechanisms I and II.

	Rate-limiting step	α_{red}
Mechanism I	31	0.5 ($\gamma_B = 0$; $n = 1$)
	32	1.5 ($\gamma_B = 1$; $n = 1$)
Mechanism II	31	0.5 ($\gamma_B = 0$; $n = 1$)
	33	2.0 ($\gamma_B = 2$; $n = 0$)

An experimental Tafel plot for the reduction of hydrogen ions (equation 30) on a tungsten surface at 300 K (under conditions in which the concentration of hydrogen ions is 2.0 mol dm^{-3}, and in the presence of an excess of potassium chloride) is shown in Figure 18. The value of i_e for the process can be determined by projecting the linear portion of the plot onto the log $|i|$ axis. The value of approximately 10^{-2} A m^{-2} is in agreement with that given in Table 1 (which it should be as the conditions are not too far removed from standard conditions). However, in this Section we are more interested in the gradient of the Tafel plot: for the plot in Figure 18, the value turns out to be −39.2 mV.

Look back at Table 2. This reveals that the value of α_{red} must be 1.5. Thus, the experimental data are consistent with Mechanism I and reaction 32 as the rate-limiting step (although of course, some other mechanism that we have not considered could be equally valid). This, then, is one example of how Tafel data can be used to elucidate reaction mechanisms. But note from the values collected in Table 3 that if the experimental data had given a Tafel slope of −120 mV (i.e. $\alpha_{red} = 0.5$ at 300 K) then we would have been unable to distinguish between Mechanisms I and II using this approach.

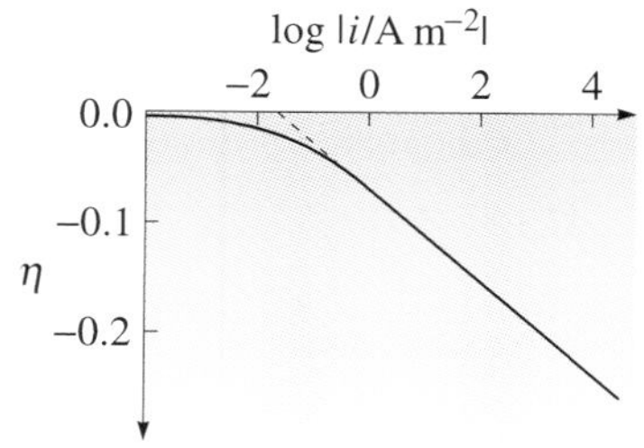

Figure 18 An experimental Tafel plot for the reduction of hydrogen ions on a tungsten surface at 300 K.

Further evidence for a proposed mechanism can be obtained in a variety of ways. For example, the effect of pH on the overpotential at a given current density could be studied, or the more normal chemical methods could be employed, such as the determination of the order of the reaction with respect to various species – H^+, for example.

Detailed investigations of the mechanism of the hydrogen ion reduction reaction in acid solution have shown that the reaction tends to proceed by:

- Mechanism I, with reaction 31 as rate-limiting, on the metals mercury, lead and cadmium;
- Mechanism I with reaction 32 as rate-limiting, on the metals nickel, tungsten and gold;
- Mechanism II, with reaction 33 as rate-limiting, on the metals platinum and rhodium.

SAQ 4 A Tafel plot for the reduction of iron(II) (Fe^{2+}) to iron metal has a slope of −120 mV at room temperature, whereas the Tafel slope for the reverse, oxidation, reaction has a value of 40 mV at room temperature.

$$Fe^{2+}(aq) + 2e = Fe(s) \qquad (34)$$

The proposed reduction mechanism is as follows:

(i) $Fe^{2+} + H_2O \longrightarrow FeOH^+ + H^+$

(ii) $FeOH^+ + e \longrightarrow FeOH$

(iii) $FeOH + H^+ + e \longrightarrow Fe + H_2O$

If the mechanism is correct, which one of the steps (i)–(iii) must be rate-limiting? (Note that equations 28 and 29 apply *only if the reaction scheme is written for overall reduction*).

Another possible mechanism is as follows:

(i) $Fe^{2+} + 2OH^- \longrightarrow Fe(OH)_2$

(ii) $Fe(OH)_2 + OH^- \longrightarrow HFeO_2^- + H_2O$

(iii) $HFeO_2^- \longrightarrow FeO + OH^-$

(iv) $FeO + H_2O + e \longrightarrow FeOH + OH^-$

(v) $FeOH + e \longrightarrow Fe + OH^-$

Can the Tafel data positively rule out this mechanism?

5.3 Tafel plots and metal extraction

Several metals are extracted from their ores by electrolysis of aqueous solutions, copper and zinc being the most important. Extraction from aqueous solution brings with it the complication of the possibility of hydrogen gas formation. This is not only energetically wasteful, but the presence of hydrogen during deposition of a metal at a cathode can give the metal undesirable qualities, such as brittleness. The aim of the industrial electrochemist is to establish conditions under which the required product is produced economically in maximum yield. At a very elementary level, our simple Tafel plots can provide some insight into the conditions necessary for successful extraction of a metal from an aqueous solution of its ions.

A Tafel plot for the reduction of M^+ ions to form a metal M, say, on an electrode composed of metal M (under standard conditions) is sketched in Figure 19. The Figure shows only the part of the Tafel plot in which we are interested, that is, for the reduction process ($\eta < 0$). The other possible reaction at the M electrode is formation of hydrogen gas from reduction of the hydrogen ions in the aqueous solution. The corresponding plot for this process *on metal M* is sketched in Figure 20.

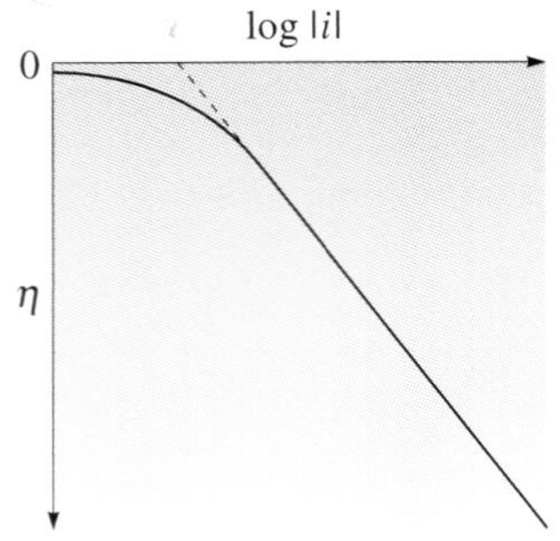

Figure 19 A sketch of a Tafel plot for the reduction of M^+ ions to metal M on an electrode composed of M.

■ If the sketches in Figures 19 and 20 are drawn to the same scale, what can be said about the relative values of i_e for the two competing processes?

■ The values of i_e are determined from the Tafel plots by projecting the linear portion of each curve onto the log | i | axis (when $\eta = 0$). Thus, Figures 19 and 20 show the situation in which the value of i_e for the reduction of hydrogen ions on metal M is less than that for the reduction of M^+ ions on metal M.

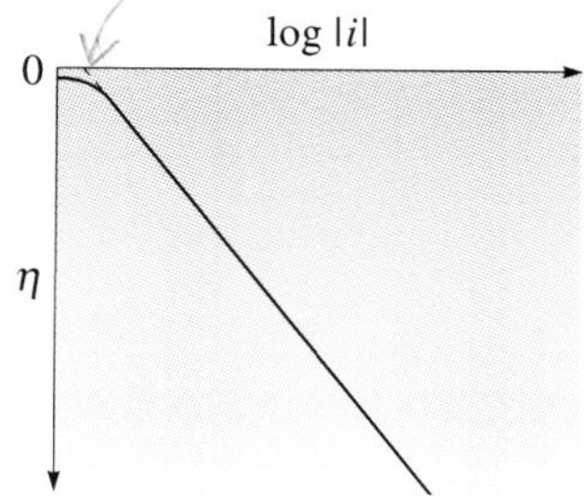

Figure 20 A sketch of a Tafel plot for the reduction of H^+ ions to hydrogen gas on metal M.

Figures 19 and 20 are also drawn assuming identical gradients for the linear portions of the plots. But how can we use the sketches in Figures 19 and 20 to decide which process is more likely under a particular set of conditions, reduction of hydrogen ions or reduction of M^+ ions? To answer this question we need to combine the sketches in some way.

■ Can we continue to label the axis of the combined plot as the overpotential? Remember that $\eta = \Delta\phi - \Delta\phi_e$, as defined in equation 13.

■ If a potential difference of $\Delta\phi$ exists across the metal–solution interface, then it is highly unlikely that the overpotential for reduction of M^+ would be the same as the overpotential for the reduction of H^+. This will occur only if the two processes have identical values of $\Delta\phi_e$.

Thus, to combine the sketches we must label the vertical axis with a term that has significance for *both* reactions. Such a term is the potential difference across the interface, $\Delta\phi$; this will be the same for both reactions. This in turn means that the two sketches will now be displaced along the vertical axis according to the difference in values of their equilibrium potential differences, $\Delta\phi_e$. Figures 19 and 20 are combined in Figure 21, where it is assumed that the concentrations remain constant at their standard values. $\Delta\phi_e(H)$, the equilibrium potential difference for the formation of hydrogen on metal M, is here assumed to be *less negative* than $\Delta\phi_e(M)$, the equilibrium potential difference for the formation of metal M on an electrode composed of metal M.

Now let's examine the situation at each of the three different values of the potential difference ($\Delta\phi_a$, $\Delta\phi_b$ and $\Delta\phi_c$) indicated in Figure 21. Well, when the potential difference across the metal–solution interface has a value $\Delta\phi_a$, *neither* the hydrogen reduction curve *nor* the metal reduction curve has been reached. Thus, at this value of the potential difference, *net oxidation* of M to form M^+ ions is taking place at the electrode. Also, there would be a tendency for H_2 gas to be oxidized to form H^+ ions in solution (but because no hydrogen gas is present in the system, this reaction cannot actually take place!). Net oxidation curves are not shown in Figure 21.

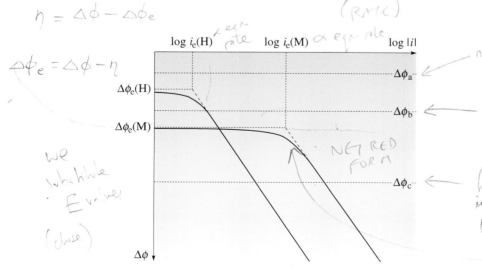

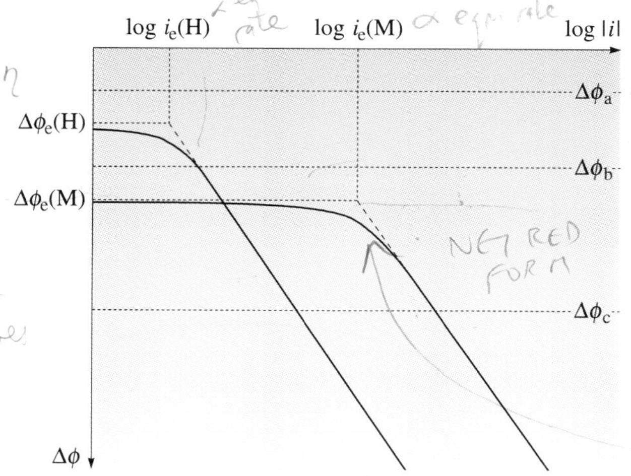

Figure 21 Schematic plots of $\Delta\phi$ versus log | i | for the competing reduction processes: $M^+(aq) + e = M(s)$ and $H^+(aq) + e = \frac{1}{2}H_2(g)$ on metal M. Here it is assumed that i_e for the reduction of hydrogen ions on metal M is less than that for the reduction of M^+ ions on metal M.

■ What can happen when the potential difference has a value of $\Delta\phi_b$?

■ At this potential difference, reduction of hydrogen ions can take place, liberating hydrogen gas at the electrode. But the potential is still more positive than the value of $\Delta\phi_e$ for metal M reduction, so oxidation of M to M^+ continues to take place.

When the potential difference has a value of $\Delta\phi_c$, however, *both* reduction processes can occur. Moreover, because of the higher value of i for the formation of metal M, the rate of formation of metal M will exceed the rate of formation of hydrogen gas. *Potentials corresponding to $\Delta\phi_c$ are the potentials used in practice for the electrolytic extraction of metals from an aqueous solution containing their ions.*

■ What will be the observed current when the potential difference has a value of $\Delta\phi_c$?

■ The *total current density* will be the current density due to the formation of metal M, i_M say, plus the current density due to the formation of hydrogen gas, i_H. Since the electrode area, A, for both processes will be the same, the observed current is given by $I = A(i_M + i_H)$.

The *efficiency* of production of metal M can be calculated from the following expression:

$$\textbf{current efficiency}\ (\%) = 100 \times \frac{i_M}{(i_M + i_H)} \qquad (35)$$

Figure 21 indicates that this efficiency will depend on the relative values of: (a) i_e for the two processes; (b) $\Delta\phi_e$ for the two processes; and (c) the Tafel slopes for the two processes.

With identical Tafel slopes (as shown in Figure 21), the current efficiency is constant *once the potential difference is more negative than* $\Delta\phi_e$(M). However, the efficiency *can* be changed by altering the experimental conditions. For example, $\Delta\phi_e$ for hydrogen formation can be changed by changing the pH of the solution.

■ According to the Nernst equation, what effect will decreasing the pH of the solution have on the value of the *electrode potential* for the hydrogen electrode? (Refer back to Block 7, Section 6.3, SAQ 23 if necessary.)

■ From the Nernst equation, $E\,(H^+|H_2)/V = -0.0592$ pH at 298.15 K. Thus, decreasing the pH of the solution will increase (that is, make less negative) the value of the electrode potential.

If the electrode potential becomes less negative, it's likely that the equilibrium potential difference, $\Delta\phi_e$(H), will also become less negative, so shifting the curve shown in Figure 21 to more positive values of $\Delta\phi$. Conversely, an increase in the pH will shift the curve to more negative values of $\Delta\phi$, thereby increasing the efficiency of production of metal M in this particular example. Changes in concentration also change the value of i_e as discussed in Section 5.1: a decrease in hydrogen ion concentration (increase in pH) *decreases* the value of i_e(H). This would shift the Tafel plot in Figure 21 to the left, further increasing the efficiency of production of metal M in this particular example.

Of course, Figure 21 represents a very simple situation: usually the plots have *different* Tafel slopes, and at higher values of the overpotential the linear relationship breaks down (recall Figure 8). In addition, the Tafel plots are often determined under standard conditions of concentration, whereas electrolysis generally takes place under conditions far from standard. Correcting the curves for such changes in concentration is not straightforward.

A further inaccuracy creeps in when we construct this sort of diagram for *real* systems; the two Tafel plots in Figure 21 are separated by the difference in their $\Delta\phi_e$ values. But we know that such values cannot be determined. However, the error introduced by equating the value of the equilibrium potential difference for a system with the *electrode potential* for that system *under the same conditions* is likely to be small, and this approach is adopted in comparing Tafel plots. Thus, even though we have said up till now that equating $\Delta\phi_e$ with E is strictly taboo (Section 2.2), we shall find that the behaviour of most electrochemical processes *can* be predicted or rationalized using this simple approximation. Nevertheless, you should always remember the distinction between the two quantities, and the approximation that is being made. If our theoretical interpretation does not match what happens in practice, then this is one area where further investigation might be fruitful.

Let's now look at some real systems. A quick glance through Section 3 of the S342 *Data Book* indicates that the standard electrode potentials of many $M^{n+}|M$ couples are more negative than that for the $H^+|H_2$ system. Thus, there are many situations that resemble the one presented in Figure 21 – under standard conditions at least. As already discussed, the separation of the two Tafel plots can often be altered by changing the pH of the solution, and pH manipulation is an important preoccupation of industrial electrochemists.

Consider the electrolytic extraction of cadmium from aqueous solution under standard conditions. Unfortunately our use of the term 'standard' now becomes a little fuzzy, because $E^{\ominus}$ values are defined at unit *activity* (a = 1.0), whereas the i_e values in Table 1 are quoted for unit *concentration* (1 mol dm^{-3}). As we stressed in Block 7, equating unit activity with unit concentration (or strictly, setting $a = c/c^{\ominus}$) involves the implicit assumption that the solutions involved show ideal behaviour. In practice, electrolyte solutions as concentrated as 1 mol dm^{-3} usually show marked 'deviations' from ideality (that is, the activity coefficient $\gamma_{\pm} \neq 1$). For simplicity, we shall use $E^{\ominus}$ values to discuss systems under standard 'concentration' conditions, but you should bear in mind that this introduces yet another approximation.

To return to the electrolysis of cadmium solutions, reference to the S342 *Data Book* provides the required $E^{\ominus}$ values: $E^{\ominus}(Cd^{2+}|Cd) = -0.40$ V and $E^{\ominus}(H^+|H_2) = 0.0$ V. Table 1 gives the relevant i_e values as 10^3 A m^{-2} for the formation of cadmium on a cadmium surface, and 10^{-7} A m^{-2} for the formation of hydrogen gas on a cadmium surface. The previous section suggested a mechanism for the hydrogen ion reduction reaction on cadmium, yielding an α_{red} value of 0.5, i.e. a Tafel slope of −120 mV. In the absence of any information on the Tafel slope for the formation of cadmium on a cadmium surface, we assume for the moment that the Tafel slopes for the two processes are identical, at −120 mV. We are now able to construct Figure 22. Notice that with the approximation being used, the vertical axis in Figure 22 is labelled V (relative to the S.H.E.) rather than $\Delta\phi$.

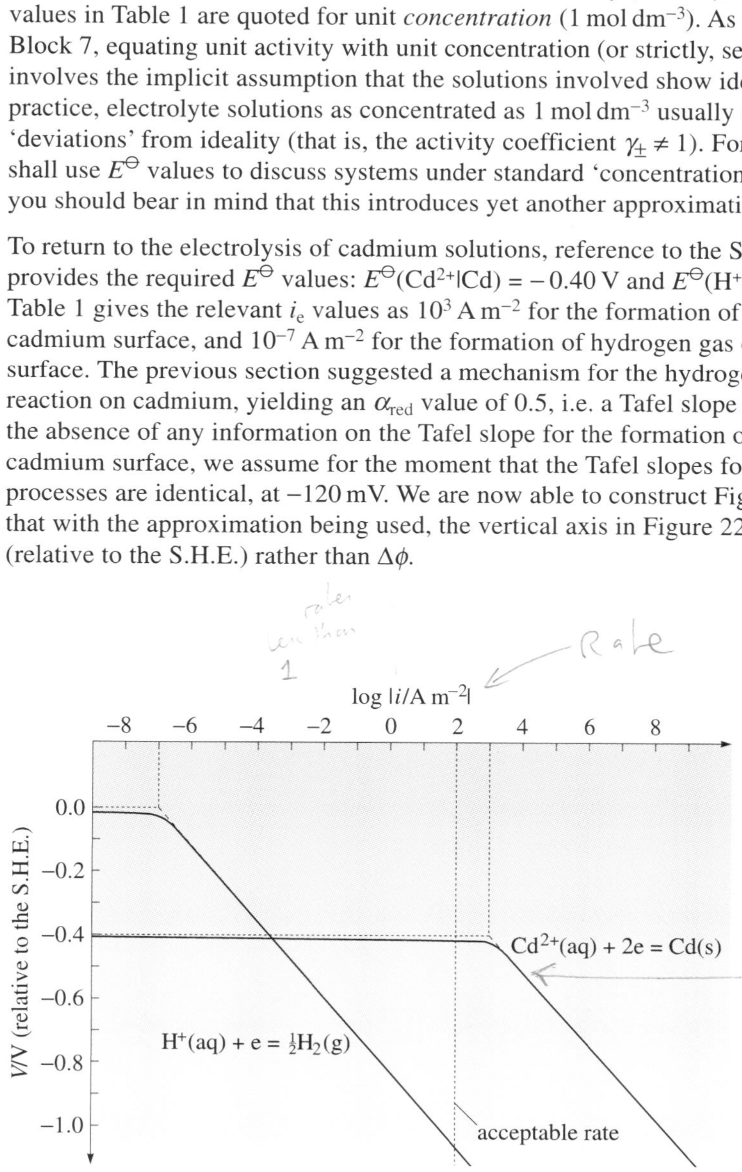

Figure 22 Plots of V (relative to the S.H.E.) versus log$|i|$ for the competing processes of hydrogen liberation and cadmium metal deposition at a cadmium electrode, assuming standard concentration conditions and identical Tafel slopes of −120 mV.

Figure 22 reveals that at potential differences more negative than – 0.40 V (relative to the S.H.E.) the rate of cadmium production will far exceed the rate of hydrogen gas formation: indeed, an acceptable rate of electrolysis (normally stated as a requirement that the current density be at least 100 A m^{-2}) will be obtained using a potential difference only marginally more negative than – 0.40 V (relative to the S.H.E.). In fact, electrolysis of cadmium solutions is usually conducted under conditions close to those we have assumed, that is, with the concentration of cadmium ions approaching 1.0 mol dm^{-3} and pH ≈ 1–2. So Figure 22 should represent something like the situation obtaining in practice. In this example, ignorance of the value of the Tafel slope for the formation of cadmium on a cadmium surface is not important since this slope doesn't feature prominently in the discussion!

As a final example, let's return to the electrolytic extraction of zinc. At first sight, this example appears similar to the cadmium example above: the $E^{\ominus}$ value for $Zn^{2+}|Zn$ is more negative than that for the $H^{+}|H_2$ couple, the i_e value for formation of zinc on zinc metal is higher than the i_e value for the formation of hydrogen gas on zinc metal (Table 1) and, indeed, electrolysis is conducted under conditions close to standard concentration conditions. But now, knowledge of both Tafel slopes is important. The Tafel slope for the formation of hydrogen on a zinc surface is known to be –120 mV, whereas that for the formation of zinc on a zinc surface is –40 mV. These data allow us to construct Figure 23, which reveals that the current efficiency for the production of zinc can be increased by making the applied potential difference (relative to the S.H.E.) more negative.

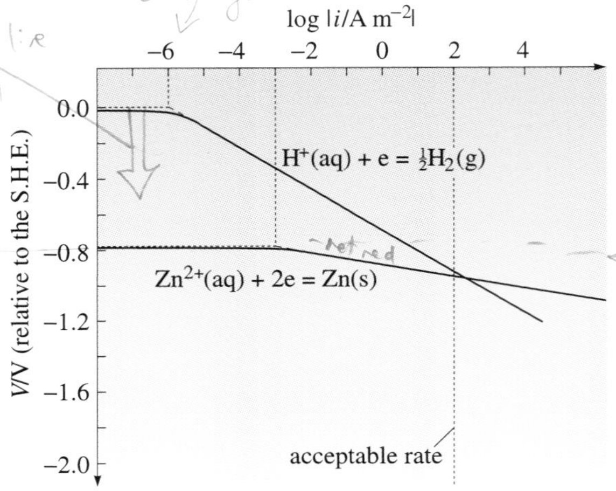

Figure 23 Plots of V (relative to the S.H.E.) versus log $|i|$ for the competing processes of hydrogen liberation and zinc metal deposition at a zinc electrode, assuming standard concentration conditions.

■ How else could the current efficiency for the production of zinc be increased?

■ The current efficiency for the production of zinc could be increased by increasing the pH of the solution. As we noted earlier, increasing the pH will shift the hydrogen ion reduction reaction to more negative values of potential difference, but will also reduce the value of i_e for this reaction. (More alarmingly, changing the pH could also change the value of i_e for the zinc reduction reaction; we take this up in Topic Study 3.)

Of course, the electrolysis of an aqueous solution is not the only way of extracting a metal electrolytically, although it is usually the cheapest method. If the problems of aqueous extraction are insuperable, then electrolysis of a molten salt is often a possibility. Aluminium, sodium, lithium and magnesium are all extracted electrolytically in this way on a fairly large scale.

SAQ 5 Nickel metal is successfully extracted by electrolysis of an aqueous solution (in which $a(Ni^{2+})$ = 1.0 and pH = 4.0) using a nickel cathode. By comparing possible Tafel plots for hydrogen gas formation and nickel metal formation, what can you say about the relative Tafel slopes for these two processes?

Use information from the S342 *Data Book* and Table 1. Assume that: (i) for the hydrogen ion reduction reaction, an increase in hydrogen ion concentration by 10^4 increases the value of i_e by a factor of 10^2; (ii) changing the pH has no effect on the value of i_e for the nickel ion reduction reaction; (iii) T = 300 K; and (iv) the equilibrium potential differences for each reduction can be equated with the corresponding values of the electrode potential. State any further assumptions that you make.

5.4 Summary of Section 5

Section 5 deals with some possible uses of data obtained from Tafel plots: i_e and α values.

1 Values of the exchange current density, i_e, provide a measure of the rate of various electrode reactions at equilibrium and can, on their own, provide an insight into some electrochemical processes. Increasing the temperature or increasing the concentration of the species involved increases the value of i_e.

2 Transfer coefficients, α values, can be related to the mechanism of an electrochemical process through equations 28 and 29. This allows some postulated mechanisms to be discounted.

3 The combined use of α and i_e values for the various possible processes taking place at an electrode provides information on the most likely electrode processes under a given set of experimental conditions. Combining the various Tafel plots requires a knowledge of the $\Delta\phi_e$ values for the various processes but, because these values can't be determined, in most cases it is a reasonable approximation to substitute E values for $\Delta\phi_e$.

4 When the two competing processes at an electrode are the deposition of metal M and the reduction of hydrogen ions from aqueous solution, the current efficiency for the production of metal M is given by the expression

$$\text{current efficiency (\%)} = 100 \times \frac{i_{\text{M}}}{(i_{\text{M}} + i_{\text{H}})} \tag{35}$$

The current efficiency can be altered by changing the pH of the solution. Often, increasing the pH increases the current efficiency for production of metal M.

6 THE VARIATION OF CELL POTENTIAL WITH CURRENT

Earlier on, we pointed out that real electrochemical systems, and certainly those of practical importance, always consist of at least two electrode–electrolyte interfaces, coupled together to form a cell. But, before trying to understand the overall behaviour of such a system, we felt it simpler first to dismantle the cell (mentally, that is!), and to study events at a *single* interface – more or less in isolation. Throughout this treatment, the central concept has been the overpotential – the deviation of the potential difference at an interface from its equilibrium value. As you have seen, the relation between overpotential and current density depends on the kinetics of the electrode process (and ultimately on the underlying reaction mechanism), the quantitative dependence being given by the Butler–Volmer equation (equation 14 in Section 3.3). With this background in place, we are now in a position to bring together two electrode systems and hence 'reassemble', as it were, a complete cell, in order to see how its **overall cell potential** depends on the current flowing through it.

As we said in Block 7, electrochemical cells fall into two broad categories, depending on whether or not the underlying cell reaction is a spontaneous process: '**self-driving' cells**, designed to convert chemical energy into electricity (batteries and fuel cells); and '**driven**' or **electrolytic cells**, designed to produce new substances. For example, metal deposition, as discussed in Section 5.3, takes place at the cathode of an electrolytic cell. We shall deal with each in turn.

6.1 Self-driving cells

As an example, consider the simple cell sketched in Figure 24. Taking account of the platinum connecting wires, we can write the cell diagram as follows:

Pt(s)|Zn(s)|Zn²⁺(aq)|Ag⁺(aq)|Ag(s)|Pt(s)

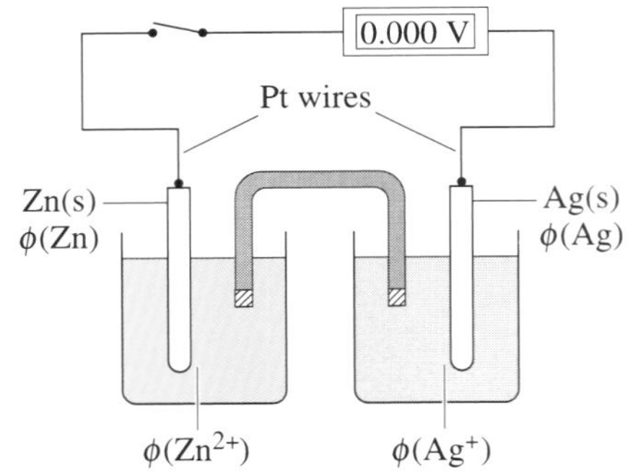

Figure 24   A simple self-driving cell.

■ Write down an expression for the potential difference V across this cell, in terms of the potential differences across *all* the interfaces involved. (Refer back to Section 2.2 if necessary.)

i.e. $\{\phi(Pt) - \phi(Ag)\} + \{\phi(Ag) - \phi(Ag^+)\} +$ *etc.*

■ 'Reading' the interfaces from right to left gives:

$$V = {}^{Pt}\Delta^{Ag}\phi + {}^{Ag}\Delta^{Ag^+}\phi + {}^{Ag^+}\Delta^{Zn^{2+}}\phi + {}^{Zn^{2+}}\Delta^{Zn}\phi + {}^{Zn}\Delta^{Pt}\phi \qquad (36)$$

where ${}^{Pt}\Delta^{Ag}\phi = \phi(Pt) - \phi(Ag)$, and so on.

■ According to equation 36, is the cell potential V likely to be independent of the current drawn from the cell?

■ No. At the very least, the potential differences ${}^{Ag}\Delta^{Ag^+}\phi$ and ${}^{Zn^{2+}}\Delta^{Zn}\phi$ are both functions of current density: this is what the Butler–Volmer equation is all about!

Before exploring this dependence further, it would be as well to say a few words about the ${}^{Ag^+}\Delta^{Zn^{2+}}\phi$ term in equation 36. It represents the potential difference between the bulk of the solution near the silver electrode, and that near the zinc (see Figure 24). In general, this will include the liquid junction potential that arises whenever two solutions of different composition are in contact. Here we shall again assume (as we did in Section 2.2) that measures have been taken – by the use of a salt bridge, say – to eliminate this effect.

$Ag^+\Delta^{Zn^{2+}}\phi = 0$

if the salt bridge is working properly, no liq junc pot

However, as soon as current is actually drawn from the cell, there will usually be a second contribution to the term ${}^{Ag^+}\Delta^{Zn^{2+}}\phi$, namely a potential drop across the *bulk* of the solution(s) between the two electrodes: we shall denote it by the symbol $\Delta\phi_S$. We ask you to accept that the *magnitude* of $\Delta\phi_S$ is given by a straightforward application of **Ohm's law** to the cell: in this case,

magnitude of potential drop across soln

$$|\Delta\phi_S| = IR_S \qquad (37)$$

where I is the current drawn from cell, and R_S is the **electrical resistance** of the solution(s). Evidently, $\Delta\phi_S$ will also make a contribution to the variation of V with I.

By recalling that the emf, E, is just the value of V when *no* current is drawn from the cell we can develop equation 38. The development is shown in Appendix 1, but attempt this for yourself if you wish.

$$V = \{E(Ag^+|Ag) - E(Zn^{2+}|Zn)\} + \eta(Ag) - \eta(Zn) + \Delta\phi_S \qquad (38)$$

where $\eta(Ag)$ and $\eta(Zn)$ are the overpotentials at the silver and zinc electrodes, respectively.

Equation 38 is, then, a general expression for the potential of the cell when current is drawn from it: we shall call this the **working cell potential**, and give it the symbol V. So what conclusions can be drawn from this simple expression? In particular (leaving aside the final term for the moment), how do the overpotentials at the two electrodes affect the cell potential?

To tackle this question, suppose for simplicity that the ions, $Ag^+(aq)$ and $Zn^{2+}(aq)$, are at unit activity. Then the electrode potentials (on the standard hydrogen scale) will have their standard values, so that, taking information from the S342 *Data Book*, at 298.15 K:

$$E_{RHE} = E^{\ominus}(Ag^+|Ag) = +0.80\ V$$

and

$$E_{LHE} = E^{\ominus}(Zn^{2+}|Zn) = -0.76\ V$$

■ According to our cell diagram, what is the *implied* cell reaction? Is this the spontaneous cell reaction under these standard conditions?

□ In line with the conventions introduced in Block 7, the implied cell reaction is

$$Zn(s) + 2Ag^+(aq) = Zn^{2+}(aq) + 2Ag(s) \quad (39)$$

Because the cell emf is positive ($E = E_{RHE} - E_{LHE} = +1.56$ V), the implied cell reaction has a spontaneous tendency to go from left to right, as written.

■ If current is now drawn from the cell, such that the overall reaction in equation 39 actually takes place, what processes will be happening at the two electrodes?

□ According to equation 39, there must be *net oxidation* at the left-hand electrode (Zn), combined with *net reduction* at the right-hand electrode (Ag).

Thus, Zn and Ag are the anode and the cathode, respectively.

Now, η(Zn) and η(Ag) are the overpotentials set up by these *net* processes at the two electrodes. As you will see in a moment, their *quantitative* dependence on the current drawn can be derived from the Butler–Volmer equation. But a *qualitative* conclusion of fundamental importance follows immediately from the conventions underlying that equation.

■ What are the *signs* of η (Zn) and η(Ag)?

□ The overpotential is positive for net oxidation (at the anode), negative for net reduction (at the cathode). Hence, $\eta(Zn) > 0$ and $\eta(Ag) < 0$.

With reference to equation 38, it follows that *both* overpotentials conspire to reduce the working potential *below* the 'thermodynamic' value given by the emf. This conclusion is represented pictorially in Figure 25, which shows how the potentials of the two electrodes (again measured on the standard hydrogen scale) 'creep' toward one another as current is drawn from the cell. This is a completely general result. Notice further that it accords, as it must, with the *thermodynamic* requirement that the maximum work, and hence maximum cell potential, is obtained when the cell is held 'in balance' and no *net* current is drawn from it.

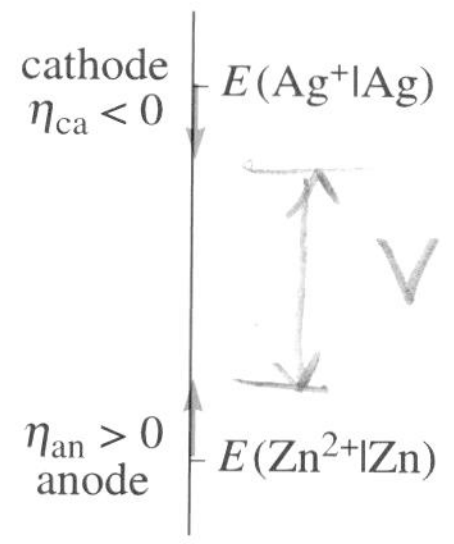

Figure 25 Drawing current from a self-driving cell reduces the cell potential below the emf.

One further point: the final term in equation 38, $\Delta\phi_S$, also tends to make the working potential of the cell *less* than its equilibrium value. To emphasize this point, we shall use equation 37 to rewrite equation 38 in the following general form:

$$V = (E_{ca} - E_{an}) + \eta_{ca} - \eta_{an} - IR_S \quad (40)$$

where E_{ca} and E_{an} are the *reduction potentials* of the half-reactions at the cathode and anode, respectively, and η_{ca} and η_{an} are the corresponding overpotentials when a given current (I) is drawn from the cell. Here, we have used the term 'reduction potential' (see Block 7, Section 6.1.2) to emphasize that E_{ca} and E_{an} are the electrode potentials of the appropriate couples *both written as reduction processes*. Thus, for the cell in Figure 24, for example, the spontaneous cell reaction (equation 39) can be 'decomposed' into the following half-reactions:

cathode (net reduction): $2Ag^+(aq) + 2e = 2Ag(s)$; $E_1 = E_{ca}$

anode (net oxidation): $Zn(s) = Zn^{2+}(aq) + 2e$; $E_2 = -E_{an}$

where E_{an} refers to the value of $E(Zn^{2+}|Zn)$.

Then,

$$E_{cell} = E_1 + E_2$$
$$= E_{ca} - E_{an}$$
$$= E(Ag^+|Ag) - E(Zn^{2+}|Zn)$$

which is the term in curly brackets in equation 38.

Given this notation, *which we shall adopt from now on*, it should be clear that:

for any self-driving cell

$$E_{cell} = E_{ca} - E_{an} > 0$$

Equation 40 is the relation between the working cell potential and current drawn that we set out to obtain: it is one of the basic equations in applied electrochemistry, and we shall make good use of it in subsequent Sections. As we hinted above, *explicit* expressions for the overpotentials at the anode and cathode can be obtained from the Butler–Volmer equation introduced in Section 3.3:

$$i = i_e \left\{ \exp\left(\frac{\alpha_{ox} F\eta}{RT}\right) - \exp\left(-\frac{\alpha_{red} F\eta}{RT}\right) \right\} \tag{14}$$

It is possible that you may be a little confused at this point. After all, equation 14 was concerned with the *competition between oxidation and reduction* at a *single* electrode. In that context, α_{ox} and α_{red} are the transfer coefficients for the competing oxidation and reduction processes, respectively: ($Ag \rightarrow Ag^+ + e$) and ($Ag^+ + e \rightarrow Ag$), say, at the right-hand electrode in Figure 24. But now we are interested in *net oxidation* at one electrode, the anode ($Zn \rightarrow Zn^{2+} + 2e$ in our example), and *net reduction* at the other, the cathode ($Ag^+ + e \rightarrow Ag$). The secret, then, is to apply equation 14 to each electrode *in turn*, under conditions that ensure this is so.

STUDY COMMENT The following SAQ gives you an opportunity to think this through for yourself: don't miss it out.

SAQ 6 What conditions guarantee that the first term in equation 14 dominates the expression for the net current density, *i*? What process will be taking place at the electrode under these circumstances? (Refer back to Section 3.3 if you are uncertain about this.) Hence show that the Butler–Volmer equation can lead to the following simple expression for the **overpotential** (for net oxidation) **at an anode**, η_{an}:

$$\eta_{an} = \left(\frac{2.303RT}{\alpha_{ox,an}F}\right) \log (i/i_{e,an}) \tag{41}$$

fragment of vine.

where $i_{e,an}$ is the exchange current density for the reaction at the anode, and $\alpha_{ox,an}$ is the corresponding transfer coefficient for the *oxidation* reaction at that electrode.

When the overpotential is reasonably large and *negative*, so that the second term in equation 14 dominates, a similar argument to that used in SAQ 6 leads to an analogous expression for the **overpotential** (for net reduction) **at a cathode**. By inspection:

$$\eta_{ca} = -\left(\frac{2.303\ RT}{\alpha_{red,ca}F}\right) \log (|\,i\,|/i_{e,ca}) \tag{42}$$

where $i_{e,ca}$ is the exchange current density for the reaction at the cathode, and $\alpha_{red,ca}$ is the corresponding transfer coefficient for the *reduction* reaction at that electrode.

We are now in a position to identify some of the factors that influence the perfomance of a self-driving cell 'under load'. Clearly, it helps if the cell reaction has a high emf. However, there is little point in selecting a system that holds great promise on this basis, if the cell potential decays dramatically the moment that current is drawn.

- Referring to equations 40, 41 and 42, how would you try to ensure that the working potential, V, remains as close as possible to the emf, E?

- The secret is to keep the terms η_{an}, η_{ca} and R_S in equation 40 as small as possible. According to equation 41 (or 42), for a given current (and hence current density, i) the overpotentials are determined by the parameters (α and i_e) that characterize the kinetics of the electrode processes; η will be small if α and i_e are large.

Figure 26 illustrates the point. Here we have taken a set of hypothetical (but not wholly unrealistic) parameters (they are listed in the caption to Figure 26), and then calculated the various contributions to V as a function of current density. For simplicity, we have assumed that the kinetic parameters are the same for the two electrode reactions (that is, $i_{e,\,an} = i_{e,\,ca}$, and $\alpha_{ox,\,an} = \alpha_{red,\,ca}$), which is unlikely to be so in practice. However, a comparison between the curves labelled A, B and C (for which R_S has the same value) serves to underline the point noted above. Notice, in particular, the beneficial effect of a high value of i_e (compare curves B and C). As noted earlier, increasing the temperature increases the value of i_e.

Apart from the thermodynamics and kinetics of the electrode processes, the other important factor in equation 40 is the resistance of the solution, R_S. Experiment shows that this depends on the **conductivity**, κ, of the solution, and on two *geometrical* parameters – the surface area (A) of the electrodes and the distance (l) between them. Formally, the relationship is as follows:

$$R_S = \frac{l}{\kappa A} \qquad (43)$$

- A comparison between curves B and D in Figure 26 underlines the importance of **minimizing the resistance of the solution**. According to equation 43, how would you try to achieve this?

- R_S would be reduced by: (a) reducing the gap between the electrodes; (b) increasing the surface area of the electrodes; and/or (c) increasing the conductivity of the solution.

In practice, (a) is largely a matter of cell design, although there can be problems with setting the electrodes too close together (shorting out, for example). Turning to point (c), the conductivity of an electrolyte solution reflects the 'efficiency' with which the ions transport current through the cell; some ions are inherently more 'mobile' than others. Without pursuing the matter further, the important point for our purposes is that

the conductivity of a *given* electrolyte solution can be increased by:

- increasing the concentration; and/or
- raising the temperature.

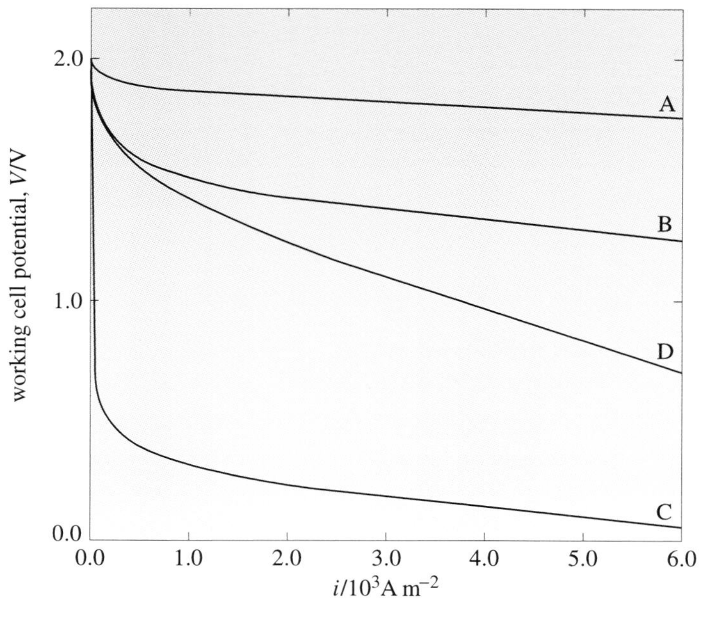

Figure 26 Cell potential versus current density for hypothetical self-driving cells with the following characteristics: $T = 300$ K; $E = 2.0$ V; electrode areas $A = 1\ \text{cm}^2$; and $i_{e,\,an} = i_{e,\,ca}$ and $\alpha_{ox,\,an} = \alpha_{red,\,ca}$ throughout.
curve A: $i_e = 10\ \text{A m}^{-2}$; $\alpha = 2.0$; $R_S = 0.1\ \Omega$;
curve B: $i_e = 10\ \text{A m}^{-2}$; $\alpha = 0.5$; $R_S = 0.1\ \Omega$;
curve C: $i_e = 10^{-4}\ \text{A m}^{-2}$; $\alpha = 0.5$; $R_S = 0.1\ \Omega$;
curve D: $i_e = 10\ \text{A m}^{-2}$; $\alpha = 0.5$; $R_S = 1.0\ \Omega$.

So we now have a second reason (in addition to increasing the rates of the charge transfer reactions at the electrodes) for running a self-driving cell at an elevated temperature.

■ The discussion above has focused on the factors that influence the terms η_{an}, η_{ca} and R_S in equation 40. Which of these factors can also affect the *emf*, *E*, of the cell?

■ The concentrations (or strictly, activities) of the species that are active in the cell reaction, and the temperature. The former affect the electrode potentials of the processes involved (via the Nernst equation), and hence the overall cell emf. The cell emf also depends on the temperature. As discussed in Block 7 (Section 7.3), the temperature-dependence of E is determined by the *sign* of $\Delta S_m^{\ominus}$ for the underlying cell reaction: the emf may either increase or decrease with increasing temperature.

In practice, the temperature-dependence of E is usually fairly small, so the beneficial effects of raising the temperature (and hence reducing η and R_S) generally outweigh any possible detrimental effect on the *maximum* cell potential.

There is one last point to note at this stage. The plots shown in Figure 26 were calculated assuming that the overpotentials at the electrodes can be expressed by equation 41 or 42, i.e. assuming that Tafel-type conditions apply, with the overpotential being neither too large nor too small. As the current drawn from the cell continues to increase then, not surprisingly, this simple relationship breaks down and V shows a catastrophic decline at a particular value of i, as shown in Figure 27.

We can explain this situation if we consider what is happening at the molecular level. Let's consider a simple reduction process. For the electrochemical reaction to take place, ions moving from the outer Helmholtz plane (OHP) towards the electrode gain an electron. The faster this reaction takes place the higher the current, but a point will be reached at which the solution cannot provide ions fast enough to the OHP and this latter process (diffusion) becomes rate-limiting. The effect is called **concentration polarization** and it leads to a **limiting current density**, i_L, as shown in Figure 27. In practice, this problem can be mitigated by stirring or agitating the solution, but this isn't easy in a conventional battery!

for reduction solution cannot provide ions quickly enough

e⁻s accumulate

DIFFUSION IS RATE LIMITER!

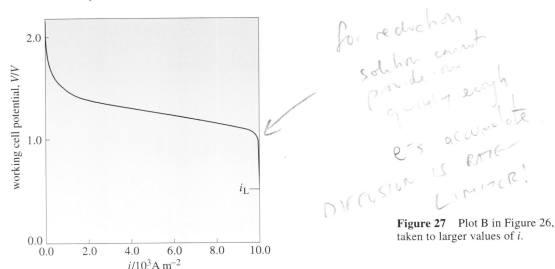

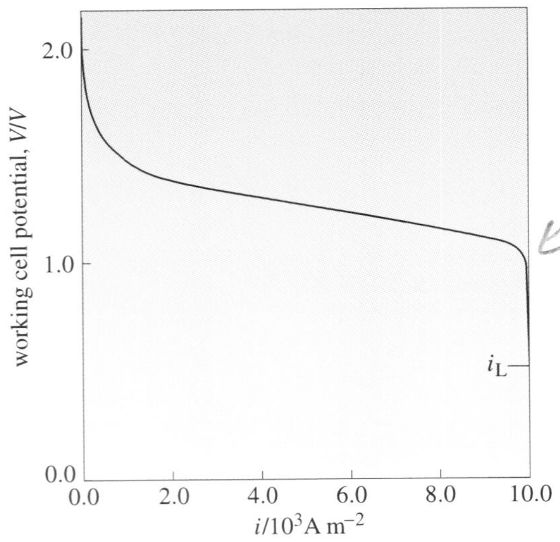

Figure 27 Plot B in Figure 26, taken to larger values of i.

6.2 Driven cells

The expression in equation 40 does, in fact, apply equally well to the potential required to 'drive' a *non-spontaneous* process at a given rate. The simplest way to see this is to return to our zinc–silver cell, only this time we include a *source* of potential in the external circuit. As indicated in Figure 28, it is connected so as to *oppose*, and then *reverse*, the spontaneous cell reaction (equation 39), the net effect being to promote the following reaction:

$$Zn^{2+}(aq) + 2Ag(s) = Zn(s) + 2Ag^+(aq) \tag{44}$$

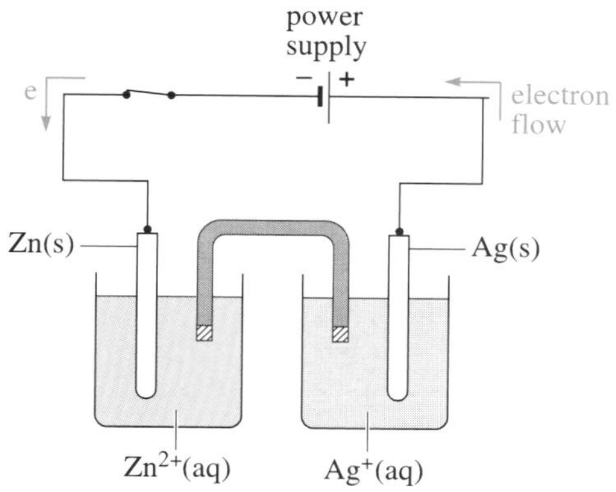

Figure 28 A simple driven or electrolytic cell.

■ In this driven cell, which electrode is the anode, and which is the cathode?

■ If the reaction in equation 44 is driven from left to right, then there must be net oxidation at the silver electrode coupled with net reduction at the zinc. Hence Ag is now the anode, and Zn is the cathode – the *reverse* of the situation discussed in Section 6.1.

■ What is the emf of the *reaction* in equation 44, under standard conditions?

■ Now, $E = E^{\ominus}(Zn^{2+}|Zn) - E^{\ominus}(Ag^+|Ag) = E_{ca} - E_{an} = -1.56$ V

As expected, $E < 0$ for this non-spontaneous process. Concentrate now on equation 40:

$$V = (E_{ca} - E_{an}) + \eta_{ca} - \eta_{an} - IR_S \qquad (40)$$

■ What do you conclude about the external potential V required to drive the reaction above at a given rate?

■ The rate of an electrochemical reaction is measured by the current flowing through the cell. With I other than zero, the terms on the right of equation 40 conspire to make *V more negative* than E.

This provides the justification for our contention (in the final Sections of Block 7) that electrolysis will occur if, and only if, the magnitude of the applied potential is *greater* than the minimum value – as given by the *magnitude* of the emf (that is, $|V| > |E|$). This conclusion is illustrated schematically in Figure 29; in this situation the electrode potentials are seen to climb *away from* one another as current is driven through the cell (compare Figure 25). As equation 40 suggests, the actual potential required, for a given rate of reaction (that is, a given current flow), depends on just the same factors as does the working potential *delivered* by the reverse, spontaneous reaction.

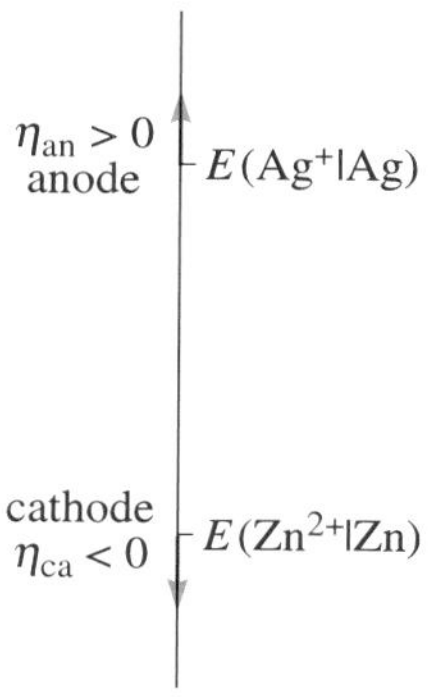

Figure 29 Passing current through a driven cell requires an applied potential greater in magnitude than the emf.

Thus, the potential required to drive a particular process can be minimized by keeping the overpotentials at the anode and the cathode as small as possible, by keeping the resistance of the solution as low as possible and by working under Tafel-type conditions to avoid problems of concentration polarization.

6.3 The general relationship between V and I

We have seen that the **relationship between current and potential for a complete cell** is given by the expression

$$V = (E_{ca} - E_{an}) + \eta_{ca} - \eta_{an} - IR_S \qquad (40)$$

where η_{ca} is less than zero (net reduction) and η_{an} is greater than zero (net oxidation). For a self-driving cell, $E_{ca} - E_{an}$ is positive so V is always smaller than E. For a driven cell, $E_{ca} - E_{an}$ is negative so V is always more negative than E. Thus the applied potential is always greater than the magnitude of the emf.

Equation 40 can be rewritten in a form that is perhaps more easily remembered:

$$V = (E_{ca} - E_{an}) - |\eta_{ca}| - |\eta_{an}| - IR_S \qquad (45)$$

Because η_{ca} is always negative, equations 40 and 45 are equivalent.

Up to this point we have implied that the resistance of the solution between the electrodes is the only resistance term that we need to consider. But this is not so, as will become apparent when we examine real cells in the next Section. Thus, the equation for V is more appropriately presented as

$$V = (E_{ca} - E_{an}) - |\eta_{ca}| - |\eta_{an}| - IR_{cell} \quad (46)$$

where R_{cell} contains contributions from terms that are a function of the design of the cell.

6.4 Summary of Section 6

The main thrust of this Section has been the development of an expression for the overall potential V of a complete cell when a current I is flowing through it. The key points are collected in Box 2, and tested in Exercise 2.

Box 2

1 The general relationship between current and potential for a complete cell is given by the equation:

$$V = (E_{ca} - E_{an}) - |\eta_{ca}| - |\eta_{an}| - IR_{cell} \quad (46)$$

with $\eta_{ca} < 0$ and $\eta_{an} > 0$.

E_{ca} and E_{an} are the electrode potentials for the couples at the cathode and anode, respectively, *both written as reductions*.

For a self-driving cell, V is always smaller (less positive) than E.

For a driven cell, V is always larger (more negative) than E.

2 If both the net oxidation reaction (at the anode) and the net reduction reaction (at the cathode) are running under high-overpotential (Tafel) conditions, the magnitude of η is:

$$|\eta| = \left(\frac{2.303RT}{\alpha F}\right) \log\left(\frac{|i|}{i_e}\right) \quad (47)$$

3 For a self-driving cell the maximum cell potential (or for a driven cell the minimum applied potential) can be obtained by keeping the overpotential terms and the resistance of the cell as low as possible. The overpotentials can be kept low by working under Tafel conditions with a system in which i_e and α are large. The resistance of the cell can be minimized by using a solution of high conductivity, at high concentration and at an elevated temperature, with electrodes that are fairly close together and have large surface areas. The resistance of the other components of the cell also needs to be considered.

STUDY COMMENT You should now attempt Exercise 2. This revises material presented in Block 7, and you might find it helpful to refer back to Section 8.1 in that Block. It also gives you your first opportunity to calculate a value of V for a real, working cell!

EXERCISE 2 Very pure hydrogen is produced on an industrial scale by the electrolysis of water:

$$H_2O(l) = H_2(g) + \tfrac{1}{2}O_2(g)$$

One type of commercial electrolyser is shown schematically in Figure 30: notice that typical working conditions include an *alkaline* electrolyte (KOH or NaOH), in place of the acidic solution assumed in Block 7.

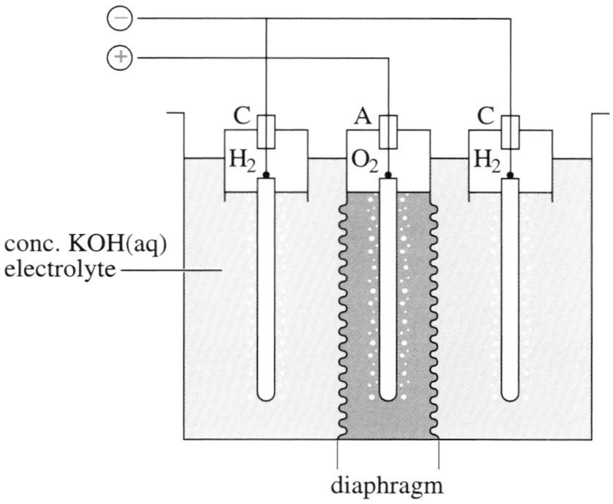

Figure 30 General arrangement of a uni-polar tank-type electrolyser for producing hydrogen: A anode; C cathode.

(a) Under these conditions, the half-reactions at the electrodes are better represented as follows (with $E^{\ominus}$ values at 298.15 K):

anode: $2OH^-(aq) = \tfrac{1}{2}O_2(g) + H_2O(l) + 2e$; $E^{\ominus}_{an} = 0.40$ V (48)

cathode: $2H_2O(l) + 2e = H_2(g) + 2OH^-(aq)$; $E^{\ominus}_{ca} = -0.83$ V (49)

rather than the more familiar forms appropriate to acidic media:

anode: $H_2O(l) = \tfrac{1}{2}O_2(g) + 2H^+(aq) + 2e$; $E^{\ominus}_{an} = 1.23$ V (50)

cathode: $2H^+(aq) + 2e = H_2(g)$; $E^{\ominus}_{ca} = 0.00$ V (51)

Convince yourself that the *standard* electrode potentials of the couples in equations 48 and 49 are identical with the values of E for reactions 50 and 51, respectively, when $a(OH^-) = 1.0$. (Remember that the values of $E^{\ominus}$ quoted above are *reduction* potentials, as noted in Section 6.1.) What is the minimum potential for the electrolysis of water at 298.15 K?

(b) In practice, the electrolysis is run at 80 °C. Taking any information you need from the S342 *Data Book*, determine the minimum potential for electrolysis at this temperature. What assumptions must you make?

(c) The type of electrolyser sketched in Figure 30 usually employs an electrolyte of 25% w/v (weight for volume) KOH (roughly 6 mol dm^{-3}). Assuming that both electrodes are nickel, use the information in Table 4 to estimate the actual potential for electrolysis at 80 °C, when the current density at each electrode is 1.5×10^3 A m^{-2}. State any further assumptions that you make. Suggest one reason why the electrolysis is run at an elevated temperature.

Table 4 Experimental Tafel slopes and exchange current densities for Exercise 2 (in 25% w/v KOH at 80 °C).

Electrode reaction	Tafel slope/mV	i_e/A m^{-2}
O_2 evolution on nickel	95	4.2×10^{-2}
H_2 evolution on nickel	−140	2.0

start with this and extrapolate to 48 and 49 using Nernst

$E_{cell} = E_{ca} - E_{an}$

0.00 − 1.23 or −0.83 − 0.40 both = −1.23 V

So you need +1.23 V at least

$\Delta G = -nFE$

$-nFE = \Delta H - T\Delta S$

7 INDUSTRIAL ELECTROLYTIC PROCESSES

7.1 Introduction

As we suggested earlier, the scope of the electrochemical industry is very broad: we have space in the closing Sections of this Block to examine just two representative examples. We begin with electrolytic processes. A major advantage of electrolytic methods is that they can be used to drive very unfavourable reactions – as in the extraction of the more electropositive metals (aluminium and sodium, for example) from their fused salts, processes that were described in the Second Level Inorganic Course.

In the less extreme cases (where an alternative purely chemical method may perhaps be available), the electrolytic route will be adopted only if it is economic to do so. In general, the decision will be based on a complex interplay between a range of factors: the cost and availability of raw materials and energy; the likely market for the product, and in particular the purity required; the capital and running costs of the plant – to name but a few. Many of these factors are common to other branches of the chemical industry. For instance, an electrode reaction is *by definition* a surface process: it occurs at the interface between electrode and electrolyte. Thus, like any other 'heterogeneous reactor', the space–time yield of an electrolytic cell will usually be an important parameter in fixing the cell design.

The main thrust, however, is to keep the applied potential as low as possible while maintaining a reasonable current so that product is formed sufficiently quickly. Thus, from our earlier discussions, ideally, we need to operate a system that will work under Tafel-type conditions, using electrodes that have high i_e and α values for the reactions of interest and in which the resistance of the cell is low. As we have seen, the concentration of the solution, the pH of the solution, the temperature, and the physical dimensions of both the cell and the electrodes are all important variables to consider. In addition, the industrial electrochemist needs to select conditions under which unwanted side reactions are suppressed (cf. Section 5.3), and possible complications due to concentration polarization are overcome through agitating, stirring or allowing the solution to flow through the cell. In practice, the choice of conditions is less straightforward than this, as we shall see.

7.2 The chlor-alkali industry: an example of 'electrochemistry in action'

The example we have chosen is the largest of the electrolytic chemical industries – the electrolysis of aqueous sodium chloride (usually brine obtained from natural salt deposits). The process sounds simple enough but in practice there are many complications. We shall start with the simple approach.

■ What products would you expect from the electrolysis of aqueous sodium chloride?

□ We need to consider all of the species present, H^+, OH^-, H_2O, Na^+ and Cl^-. The competing reactions at the cathode will be reduction of H^+ and reduction of Na^+, and at the anode they will be oxidation of OH^- and oxidation of Cl^-. Reduction of Na^+ ions to sodium metal is not possible on both thermodynamic and kinetic grounds and so need not be considered. The reactions of interest, therefore, are reduction of H^+ at the cathode to form hydrogen gas, and oxidation of OH^- and Cl^- ions at the anode.

The actual products at the electrodes are hydrogen gas at the cathode and chlorine gas at the anode; the oxidation of OH^- must be suppressed in some way. In addition, as hydrogen ions and chloride ions are being removed from the aqueous solution, what's left in solution is the other product – aqueous sodium hydroxide.

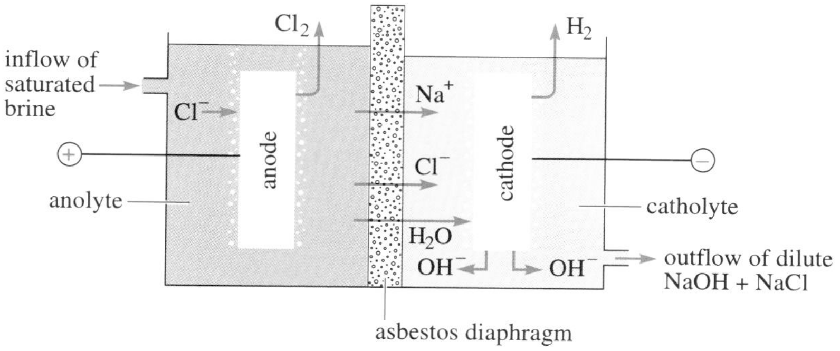

Figure 31 General arrangement of a diaphragm cell for producing chlorine and sodium hydroxide by the electrolysis of brine. A head of NaCl solution in the anode compartment keeps solution flowing from left to right through the porous diaphragm, and prevents migration of OH^- ions from right to left.

Chlorine and sodium hydroxide (caustic soda) are two of the most widely used chemicals in the world, being vital to a large sector of both the inorganic and organic (in plastics and solvents manufacture, for example) chemical industries. Witness the fact that annual production of chlorine currently (1995) runs at around 10^6 tons in the UK (about 10^7 tons in the USA – requiring some 0.2 square mile of anode, and 3.3×10^7 *mega*watt-hours of electrical energy!) The hydrogen produced is used in many chemical processes and as a fuel, often to help generate the power required for electrolysis. However, we chose this example not only because of its importance, but also because very different technologies – *diaphragm and membrane cells*, on the one hand, and *mercury cells* on the other – exist side by side. Here, we shall concentrate mainly on the first of these; the principles behind the mercury cell are taken up briefly in Exercise 3 (Section 7.2.3). One further point motivated our choice: although the first brine cells were introduced in the 1890s, recent years have seen major improvements in technology – arising, in part at least, from a better appreciation of electrode kinetics.

The basic design and modus operandi of one type of industrial cell – the *diaphragm cell* – are indicated in Figure 31. We shall have more to say later about the **separator** (be it a diaphragm as in Figure 31, or a membrane), but one of its roles is simply to keep the products apart: hydrogen and chlorine form potentially explosive mixtures!

The reaction taking place at the anode is

$$2Cl^-(aq) = Cl_2(g) + 2e \qquad (52)$$

The pH of the brine solution in this compartment – the *anolyte* – must be maintained slightly acidic in order to minimize the hydrolysis of chlorine to hypochlorite, OCl^-:

$$Cl_2(g) + OH^-(aq) = HOCl(aq) + Cl^-(aq)$$

$$HOCl(aq) = H^+(aq) + OCl^-(aq)$$

In the cathode compartment the pH of the *catholyte* is at least 14, as a result of the reaction that takes place. Under these conditions, the reaction at the cathode is *not* reduction of hydrogen ions but is more appropriately described (cf. Exercise 2) as

$$2H_2O(l) + 2e = H_2(g) + 2OH^-(aq) \qquad (49)$$

A second role of the separator, then, is to help to maintain this difference in pH between the two compartments. The diffusion of hydroxide ions from the cathode compartment to the anode compartment through the porous asbestos diaphragm is inhibited by maintaining a higher level of brine solution in the anode compartment such that the solution flows from left to right in Figure 31. The outflow from the cathode compartment is therefore an aqueous solution of sodium ions, hydroxide ions and (depleted) chloride ions.

■ Can you think of another reason for flowing the solution through the cell?

■ As well as allowing products to be continuously extracted, the flow of solution will reduce any concentration polarization effects.

Before we get engrossed in the kinetics of the processes, let's first of all consider the thermodymanic factors. The electrode potentials for the reactions of interest are listed in Table 5 for conditions that typically prevail at the anode and the cathode of industrial cells. The incoming saturated brine has a concentration in the range 140–300 g dm^{-3} (roughly 3–5 mol dm^{-3}, dependent on the temperature).

Table 5 Electrode potentials for reactions involved in brine electrolysis (all at 298.15 K).

Reaction	pH	$c(\mathrm{Cl^-})$/mol dm^{-3}	E/V
anode compartment			
(a) $Cl_2(g) + 2e = 2Cl^-(aq)$	4	4.0	1.32
(b) $\frac{1}{2}O_2(g) + 2H^+(aq) + 2e = H_2O(l)$	4	4.0	0.99
cathode compartment			
(c) $2H_2O(l) + 2e = H_2(g) + 2OH^-(aq)$	14	variable	−0.83

STUDY COMMENT SAQ 7 further revises the material in Block 7. You should attempt it if you are not yet confident when using the Nernst equation to calculate the values of electrode potentials.

SAQ 7 (revision) Use information from the S342 *Data Book* to check the entries in Table 5. (It may help to refer back to the answer to Exercise 2.) What assumptions are involved in these calculations?

- According to the information in Table 5, what is the minimum potential for chlorine production via electrolysis under industrial conditions?

- The required electrode reactions are labelled (a) and (c) in Table 5. Thus, the cell potential $E = E_{ca} - E_{an} = E(c) - E(a) = (-0.83 - 1.32)$ V = −2.15 V. Thus the minimum external potential required for electrolysis is 2.15 V.

  Concentrate now on the reactions labelled (a) and (b) in Table 5. Do you foresee a problem? (Again it may help to refer back to Exercise 2.)

The reaction labelled (b), or rather the reverse, oxidation process (equation 50 in Exercise 2), takes place at the anode during the electrolysis of water (at pH < 7, that is). The crucial point here is that its electrode potential is *smaller* (less positive) than that of the desired process – the oxidation of chloride. Thus, on thermodynamic grounds, it seems that oxygen (from oxidation of the solvent) rather than chlorine will be the favoured product at the anode. At the very least, there is a need to *suppress* oxygen evolution, in order to keep the current efficiency for chlorine production as high as possible (cf. the discussion in Section 5.3). This is where the choice of electrode material becomes crucial.

7.2.1 The cathodic reaction

The reaction of interest at the cathode is the formation of hydrogen gas.

- From our earlier discussion, and from the data in Table 1, which material would you select to use as the cathode?

- From Table 1, the best electrocatalysts for the hydrogen evolution reaction are the noble metals platinum and palladium. In addition, the discussion in Section 5.2 revealed that the α_{red} value for this reaction on such metals is of the order of 2.0. But this is for acidic conditions: the reaction we are interested in here takes place at pH 14!

In practice, it turns out that these noble metals *are* the best metallic electrocatalysts for the hydrogen evolution reaction, even under the highly alkaline conditions of interest to the chlor-alkali industry.

- Use equation 47 to estimate the magnitude of the overpotential for production of H_2 gas (the so-called **hydrogen overpotential**) on a platinum electrode at 298.15 K and at a typical working current density of 2×10^3 A m^{-2}. In strong alkali (pH 13–14), i_e is about 4 A m^{-2} at 298.15 K, and the Tafel slope is around 100 mV.

- $|\eta| = (100\ \mathrm{mV}) \log\{(2 \times 10^3)/4\} \approx 270\ \mathrm{mV}$

■ Can you suggest ways in which the overpotential on platinum could be reduced?

■ The overpotential is a function of both the concentration of hydrogen ions and the temperature. If the pH of the solution is fixed then the value of η could be reduced by raising the temperature. Raising the temperature increases the value of i_e, which in turn reduces the value of η.* Thus, commercial electrolysers are usually run at around 60–70 °C.

On a more practical note, however, it is clear that the best metal electrocatalysts for this reaction – and indeed for many others – are also the most expensive! Until recently, therefore, the cathode actually used in diaphragm and membrane cells has been a low-carbon steel, for which the hydrogen overpotential is typically around 400 mV under working conditions. However, *coatings* of catalytic nickel alloys are now available, which decrease this value to 150–200 mV. There are high expectations that improvements in such coatings will reduce it yet further, perhaps to a figure as low as 20–50 mV. Figure 32 shows typical experimental data for one such coating under conditions that prevail at the cathode of a diaphragm cell.

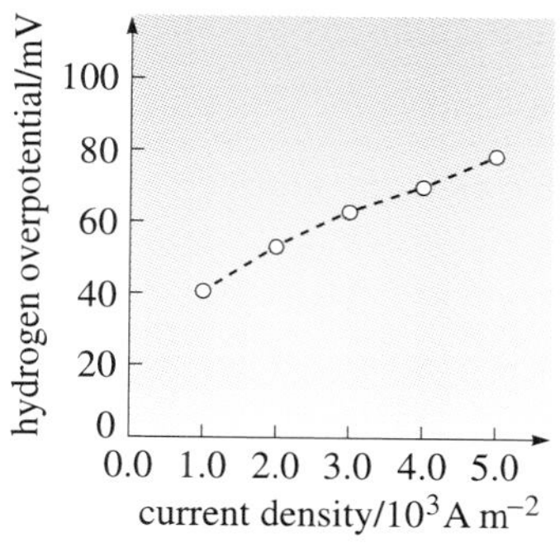

Figure 32 The overpotential for hydrogen evolution at a steel electrode coated with a Ni–Mo alloy developed by BP, as a function of current density. (17% NaOH−15% NaCl aqueous electrolyte; temperature = 70 °C.) The dashed lines have no significance other than indicating the general trend.

These coatings are prepared in such a way that they have large surface areas, which are often difficult to measure. This can lead to problems in interpretation: the reduction in overpotential could be due as much to an underestimate of the area as to an increase in the value of i_e.

7.2.2 The anodic reaction

By this stage, you probably have an inkling that the requirement for the anode material is even more stringent: it must *selectively* catalyse the evolution of chlorine, but *not* that of oxygen. An additional factor is the high concentration of chloride ions in the anode compartment, which (as you will see in Topic Study 3) makes this a highly corrosive environment: the availablity of a sufficiently stable (and economical) anode was therefore a prerequisite for the growth of the chlor-alkali industry.

For most of the history of the industry, synthetic graphite or some related form of carbon was used as the anode – even though, with an overpotential for chlorine evolution of around 500 mV under operating conditions, this could hardly be described as a 'good catalyst'. Such anodes also had a service life of only 6–24 months, being consumed at the rate of 1–3 kg per tonne of chlorine produced – owing in part to direct reaction with any oxygen evolved to give CO_2.

These deficiencies were well recognized, and there was some attempt to introduce 'valve metal' (usually titanium) electrodes, coated with precious metals (Pt or Pt–Ir alloys, for example) in dispersed form, the idea being to combine the conductivity and corrosion-resistance of the former with the catalytic power of the latter. The reduction in overpotential was certainly significant – to around 100 mV for chlorine production – but electrodes were expensive, and the loss of precious metal in service was unacceptably high.

A major advance for the industry came in the 1960s with the development (by H.B. Beer, an independent Dutch inventor, among others) of materials now known collectively as **dimensionally stable anodes, DSAs** (so-called because they have long lives, and are not subject to a disastrous loss of the catalytic coating if the cell shorts out for any reason). Although the exact composition of commercial DSAs is a closely guarded trade secret, they are all titanium-based with a coating of ruthenium dioxide (RuO_2) containing various amounts of other transition-metal oxides, such as Co_3O_4. Again, it is the coating that imparts catalytic activity: the effect is dramatic indeed, with the chlorine overpotential on commercial DSAs being around 40 mV.

A major problem with these oxide electrodes has been to develop formulations that do not also promote the oxidation of water, which leads to contamination of the chlorine by oxygen. (This should come as no surprise, in view of our discussion of

* Of course, changing the temperature also changes the *minimum* potential for electrolysis, but the effect is (as usual) fairly small. In this case, $E(343\ \text{K}) = -2.13\ \text{V}$ (compare $E(298.15\ \text{K}) = -2.15\ \text{V}$), as you can confirm for yourself.

transition-metal oxides as *non*-electrochemical oxidation catalysts in Block 5.) Thus, for example, the Mond 1 coating developed by ICI is essentially a mixture of RuO_2 and TiO_2. Whereas the kinetic parameters for chlorine evolution are effectively independent of the RuO_2 content of the coating, Figure 33 shows that, over the range 30–90 mass %, the oxygen overpotential falls quite dramatically as the proportion of RuO_2 increases.

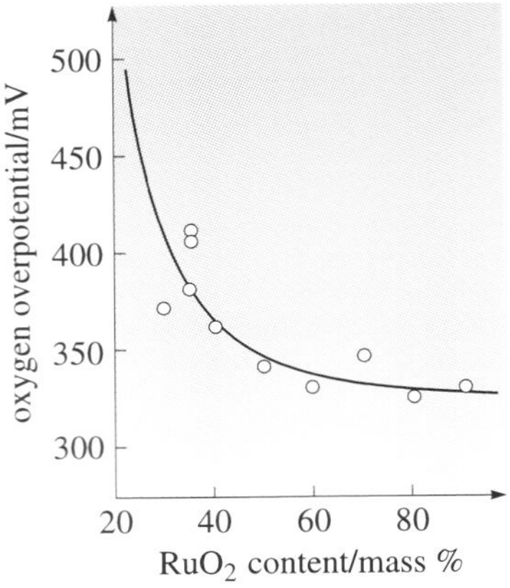

Figure 33 Effect of RuO_2 content in the Mond 1 anode coating on the overpotential for oxygen evolution from aqueous solution. (1 mol dm^{-3} H_2SO_4 electrolyte; temperature = 30 °C; current density = 1.5×10^3 A m^{-2}.)

Further problems stem from the fact that coatings like Mond 1 are polycrystalline and complex in structure. Not surprisingly, their electrical and electrocatalytic properties are found to be highly sensitive, both to their method of preparation and to the electrolysis conditions. Particularly marked is the effect of changing pH on the kinetics of chlorine evolution at such electrodes: sample results for a Mond-type coating are shown in Figure 34. Here (these results are quite typical) the Tafel slope for chlorine evolution is seen to increase with increasing pH. This effect has been attributed, in part at least, to alterations in the surface composition and/or structure of the coating brought about by changes in pH. But whatever the underlying reason, the practical message of Figure 34 is clear – it is important to maintain a low pH in the anode compartment of the cell to minimize the overpotential.

This requirement is sharpened by the fact that the rate of the competing reaction at the anode – oxygen evolution – is also pH-dependent: the higher the pH, the higher the rate. As with chlorine evolution, surface changes to the oxide coating undoubtedly play their part, so there is probably a catalytic element to this dependence. But now there is a second, more important factor. It is seen most simply by rewriting the equation in the form more appropriate to alkaline conditions (Exercise 2):

$$2OH^-(aq) = \tfrac{1}{2}O_2(g) + H_2O(l) + 2e \qquad (48)$$

Evidently, raising the pH increases the concentration of reactant – OH^- ions – for this process, and thus enhances the rate.

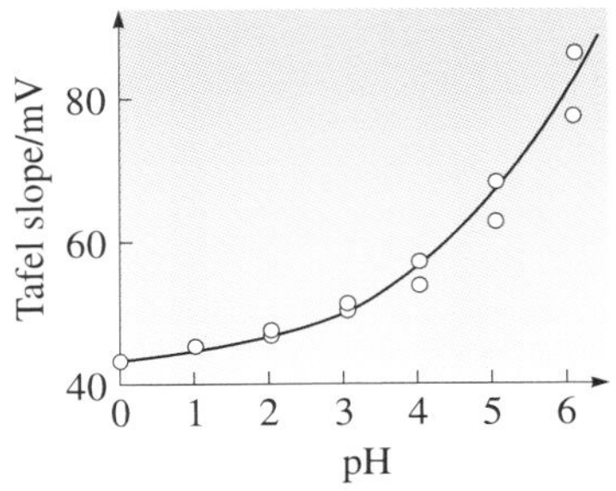

Figure 34 Tafel slope as a function of pH for chlorine evolution on a RuO_2/TiO_2 electrode, from a Cl_2-saturated NaCl–HCl aqueous solution with $c(Cl^-) = 5$ mol dm^{-3}, and at a temperature of 25 °C.

- What reduction in the energy requirement does this represent, as a percentage of the minimum thermodynamic requirement?

- According to the discussion above, the reduction is at least 0.46 V (500 mV to 40 mV) in 2.15 V, that is, about 20%.

It was this development, perhaps more than any other, that enabled the industry to cope with the successive 'energy crises' of the 1970s. (See also Table 6 in the next Section.)

7.2.3 The cell design

According to equation 46, the other major factor that determines the energy consumption of the chlor-alkali process is the internal resistance of the cell, R_{cell}: this must be kept to a minimum. Take the diaphragm cell (Figure 31), for example: the basic design features of a real industrial cell are sketched in Figure 35. Notice the flow of electrolyte through the cell, and the fact that the diaphragm (based on asbestos, with various polymers added to improve its performance) is actually attached to the cathode(s).

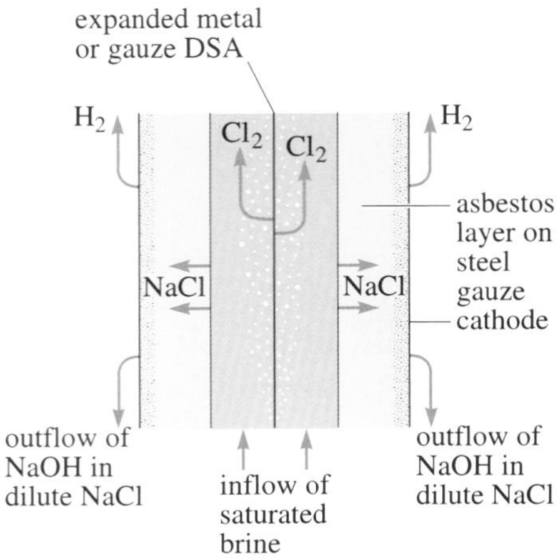

Figure 35 Principle behind the design of a diaphragm cell.

The resistance of the solution is minimized by using a high concentration of brine (usually, saturated brine is used) at an elevated temperature and with electrodes of large surface area and a small gap between them. By constructing the cell such that the diaphragm is attached to the surface of the cathode, the resistance of the solution is restricted to the resistance of the anolyte. The resistance of the cell is further reduced by making the electrodes of gauze or expanded metal. This is basically to allow rapid release of the bubbles of gas formed (hydrogen and chlorine). The upward flow of electrolyte also helps to sweep the anode surface free of bubbles (and reduces concentration polarization effects).

But there are several problems with the diaphragm. As the information collected in Table 6 reveals, despite improvements in the composition of the diaphragm (compare cells 1 and 2 with cells 3 and 4), it still makes a substantial contribution to the cell resistance. We discussed a further problem earlier. An asbestos separator acts simply as a physical barrier: its porous structure allows the necessary ion transport to take place, but it cannot *discriminate* between species. Thus, the sodium hydroxide solution drawn off from the cathode compartment is always contaminated with chloride ions. Potentially more damaging, however, would be the movement of hydroxide ions in the opposite direction, which would *raise* the pH of the anolyte. This would have three undesirable effects: it would reduce the rate of chlorine evolution, while enhancing that of the competing process, and it would also promote the hydrolysis of chlorine to hypochlorite.

Table 6 Potential distribution in diaphragm cells (current density at the electrodes = 2.17×10^3 A m^{-2}; $E \approx 2.2$ V).[a]

Cell features	$\lvert V_{ca} - V_{an}\rvert$/V	IR_S/V	$IR_{diaphragm}$/V	Total/V
1 Graphite anode; steel cathode; standard asbestos diaphragm (1968)	2.68	0.87	0.66	4.21
2 DSA box anode; steel cathode; standard asbestos diaphragm	2.47	0.44	0.66	3.57
3 DSA expanded anode; steel cathode; modified diaphragm	2.47	0.32	0.37	3.16
4 DSA expanded anode; Ni-coated cathode; modified diaphragm	2.31	0.32	0.37	3.00

[a] Data taken from an article by T.A. Liederbach in *Diaphragm Cells for Chlorine Production* (1977), Society of Chemical Industry, London. The term $\lvert V_{ca} - V_{an}\rvert$ includes *both* the thermodynamic potential *and* the overpotential at each electrode. (Notice the improvement achieved by introducing first an anode catalyst, and then a cathode catalyst.) Cell potentials recorded in the industrial literature always include an allowance for the resistance of the *external* electrical circuit, apportioned between the banks of cells in the cell house. Here this amounts to an extra 0.21 V in each case.

The net effect is to constrain the operating conditions for diaphragm cells: in practice, the concentration of hydroxide ion formed at the cathode must not be allowed to rise above around 12%. Because sodium hydroxide (an equally important product of this industry) is normally traded as a 50% solution, this in turn necessitates the inclusion of a second, energy-intensive, step in which water is removed by evaporation. Moreover, the product still contains about 1% sodium chloride as an impurity: although processes to remove this salt do exist, they are relatively complex and expensive to operate.

The new generation of **membrane cells** (Figure 36) already goes a long way toward overcoming these problems: here, the separator is a **cation-exchange membrane**. Notice that the anolyte and the catholyte streams are now quite separate, and there is no *bulk* flow of electrolyte through the cell. On the contrary, the object is to design a membrane that permits *only* Na^+ ions to pass, as indicated in Figure 36. Recent years have seen rapid progress in this area: all the membranes developed to date are perfluorinated polymers, with side-chains including sulfonic acid or carboxylic acid groups (Figure 37), or both. In use, they take the form of thin films, usually reinforced by a plastic net. Membranes are available that allow concentrations of sodium hydroxide as high as 35% to be produced *directly* – before, that is, transport of OH^- ions through the membrane becomes a significant problem. An additional spin-off is a noticeable reduction in the cell resistance, allowing present membrane cells to run at current densities roughly twice those for diaphragm cells, without a significant penalty in raising the required potential.

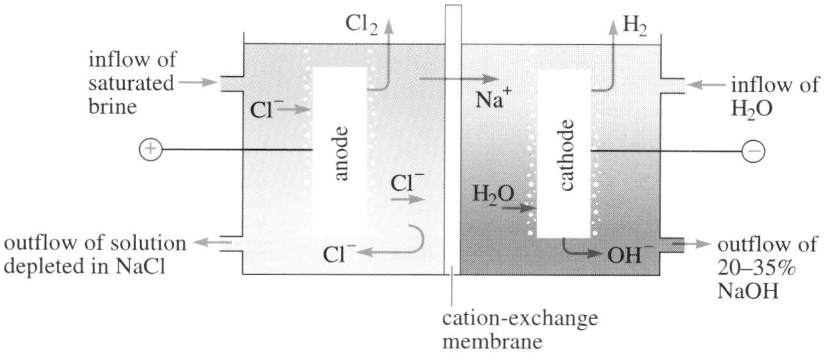

Figure 36 General arrangement of a membrane cell for brine electrolysis. Ideally, the membrane only allows passage of positive ions: OH^- ions cannot get into the anode compartment, and chloride ions cannot get into the cathode compartment to contaminate the sodium hydroxide solution.

(a) $(CF_2\!-\!CF_2\!-\!CF\!-\!CF_2)_x$
                     |
                  $(OCF_2\!-\!CF)_y\!-\!OCF_2CF_2\!-\!SO_2OH$
                           |
                         CF_3

(b) $(CF_2\!-\!CF_2)_x\!-\!(CF_2\!-\!CF)_y$
                              |
                         $(OCF_2\!-\!CF)_m\!-\!O(CF_2)_n\!-\!COOH$
                                 |
                               CF_3

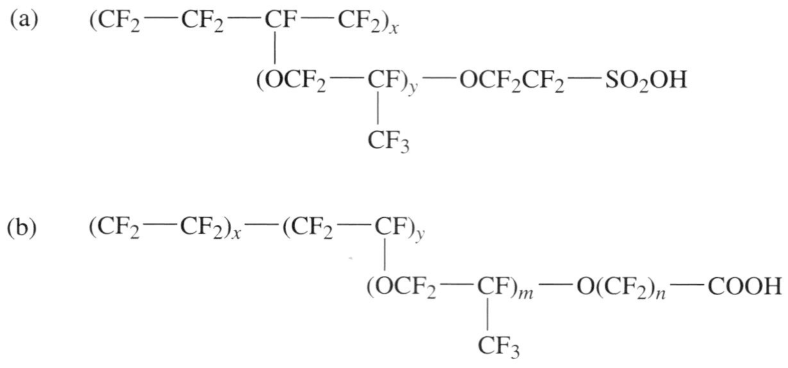

Figure 37 Perflouroplymers for ion-permeable membranes used in brine electrolysis: (a) Nafion, manufactured by E.I. du pont: (b) Flemion, manufactured by Asahi Glass Co. Ltd.

The third, and much older, type of technology is the **mercury cell**; a typical example is sketched in Figure 38, and examined in Exercise 3. Here the reaction at the anode is as before, but the reaction at the cathode is not now formation of hydrogen gas: this reaction has been suppressed by the use of a mercury cathode.

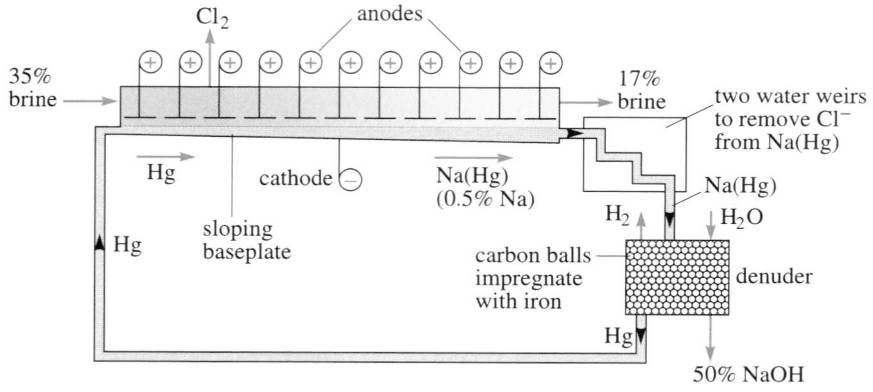

Figure 38 Representation of a mercury cell, and coupled 'circuit' for recycling mercury and producing 50% sodium hydroxide solution.

EXERCISE 3   In the mercury cell, the electrode reactions are:

anode: $2Cl^-(aq) = Cl_2(g) + 2e$ (52)

cathode: $2Na^+(aq) + 2e + 2Hg(l) = 2Na(Hg)$; $E^{\ominus}(298.15\ K) = -1.89\ V$ (53)

where Na(Hg) represents the *sodium amalgam* that is formed at the cathode. This is run off from the cell (as indicated in Figure 38), and decomposed in a separate reactor – the *denuder* – in the presence of a transition-metal catalyst (usually Fe or Ni):

$2Na(Hg) + 2H_2O(l) = H_2(g) + 2Na^+(aq) + 2OH^-(aq) + 2Hg(l)$ (54)

Thus, the *overall* process can be represented by equation 55:

$2H_2O(l) + 2Cl^-(aq) = H_2(g) + Cl_2(g) + 2OH^-(aq)$ (55)

(a) What is the minimum potential for electrolysis to produce the desired products in this cell? (Assume that the electrolyte is 4 mol dm^{-3} NaCl with a pH of 4, as in Table 5, Section 7.2.) What further assumptions are involved in this calculation?

(b) In industry, the applied potential is usually in the range from 4.1 V to 4.5 V. On thermodynamic grounds alone, would you expect H$_2$ also to be evolved at the cathode? Can you suggest why this is not a problem in practice? What information would you require in order to estimate the relative rates of the competing processes at the cathode?

(c) Do you now see a conflict between your answer to part (b), and the fact that the reaction in equation 54 occurs rapidly in the denuder? Try to suggest a plausible explanation.

(d) Would you expect the internal resistance of a mercury cell to be higher or lower than that of a typical diaphragm or membrane cell?

7.2.4 A comparison of chlor-alkali cell types

To draw together our discussion of the chlor-alkali industry, consider the information on the three different types of cell under typical operating conditions shown in Table 7. All are relatively modern cells using DSAs (or their equivalent). Notice that all three cells operate with similar current efficiencies, and that the mercury cell operates at a higher potential and higher current density than the other two cells, so that a given amount of product can be formed in a shorter time. The energy consumption for each process, given as the amount of energy required to produce one ton of sodium hydroxide solution, is seen to be lowest for the membrane cell and highest for the mercury cell, but remember that the concentration of the sodium hydroxide solution produced in the mercury cell is much higher than that produced in the other two cells. If the energy requirement to produce 50% sodium hydroxide solution is included (bottom row of Table 7) then the mercury cell is seen to have a slight energy advantage over the diaphragm cell, the membrane cell still being the most energy efficient.

Table 7 Typical data for recent commercial chlor-alkali cells.[a]

	Mercury	Diaphragm	Membrane
operating potential/V	−4.4	−3.45	−2.95
current density/10^3 A m^{-2}	10.0	2.0	4.0
current efficiency for Cl_2/%	97	96	98.5
energy consumption (kWh/ton of NaOH solution)			
electrolysis only	3 150	2 550	≈2 400
electrolysis + evaporation to 50% NaOH	3 150	3 260	≈2 520

[a] Data taken from D. Pletcher and F. Walsh, *Industrial Electrochemistry*, Second edition (1990), Blackie Academic and Professional.

But the energy requirement is not the only factor to take into account: the capital cost of mercury is extremely high, and considerable ancillary plant is required to remove it from product and effluent streams. Indeed, environmental considerations (especially in Japan, where the industry was faced with government action on mercury pollution) were a powerful force in the development of membrane technology during the 1970s. Membrane cells have a lower energy requirement, they are much easier to operate and maintain and, in addition, appear to offer low capital and maintenance costs – with the exception of the membranes themselves, which are not cheap.

7.2.5 Summary of Section 7.2

The chlor-alkali industry is concerned with the production of chlorine and sodium hydroxide solution by the electrolysis of brine. Three different technologies have been used: diaphragm cells, membrane cells and mercury cells.

1 Diaphragm and membrane cells are similar in principle. In both cases, the electrode processes are as follows:

cathode: $2H_2O(l) + 2e = H_2(g) + 2OH^-(aq)$ (49)

anode: $2Cl^-(aq) = Cl_2(g) + 2e$ (52)

The separator in these cells has two key roles:

- to keep the gaseous products (H_2 and Cl_2) apart; and
- to inhibit the hydrolysis of Cl_2 to hypochlorite by maintaining a low pH in the anode compartment.

(a) *In diaphragm cells*, the separator acts as a simple porous barrier: migration of OH^- ions into the anode compartment is inhibited by brine flowing through the cell (see Figure 31). The solution from the cathode compartment is effectively dilute sodium hydroxide, but it is always contaminated with chloride ions.

(b) *In membrane cells*, the diaphragm is replaced by a cation-exchange membrane, which permits the passage of only positive ions. There is no *bulk* flow of electrolyte through the cell: rather, brine is cycled through the anode compartment, and the cathode compartment is kept topped up with water (see Figure 36). This allows a more concentrated NaOH solution, uncontaminated by NaCl, to be produced directly.

2 The distinctive feature of a mercury cell is the use of flowing mercury as the cathode. Electrolysis produces Cl_2 (at the anode, reaction 52), and a sodium amalgam (at the cathode), which is run off from the cell. Reaction with water in a separate reactor yields a 50% NaOH solution, uncontaminated with NaCl. Although they are more energy efficient overall than diapraghm cells, concern about the toxic nature of mercury has made this type of cell less competitive.

3 Other advances in the chlor-alkali industry reflect the drive to keep the applied potential (V) to a minimum, while generating products at an acceptable rate. With reference to our general expression for V,

$$V = (E_{ca} - E_{an}) - |\eta_{ca}| - |\eta_{an}| - IR_{cell} \qquad (46)$$

the main points can be summarized under three headings.

Electrode kinetics

Keeping the overpotential terms to a minimum requires electrode processes with high i_e and α values. There are two main strategies:

(a) Cells are run at an elevated temperature (typically 60–70 °C), which increases i_e.

(b) The development of electrode materials that promote the desired processes (e.g. equations 49 and 52), but suppress unwanted reactions (specifically, O_2 evolution at the anode). In practice:

- Modern cells of all types now use dimensionally stable anodes (DSAs) – titanium-based materials with a coating of RuO_2 containing various amounts of other transition metal oxides. By replacing graphite anodes with DSAs, the industry achieved a striking reduction in η_{an} (and improved the current efficiency for Cl_2 production.)
- In diaphragm and membrane cells, Pt and Pd are the best electrocatalysts for H_2 evolution at the cathode, but they are expensive. Steel coated with nickel alloys has proved to be an effective (and cheaper) alternative.

Concentration polarization effects

Such effects are minimized in all types of cell by flowing the solution through the cell.

The cell resistance, R_{cell}

(a) The solution resistance is minimized by using a high concentration of NaCl at an elevated temperature, with large-surface-area electrodes set close together.

(b) The development of better diaphragms and (with the prospect of further improvements yet to come) ion-exchange membranes has reduced the separator's contribution to R_{cell}.

4 The developments summarized above coincided with a period of rapidly escalating energy costs; their ready acceptance by the industry undoubtedly owes much to this pressure.

7.3 Electrosynthesis: a few general remarks

Having described the chlor-alkali process in some detail, we shall not attempt a review, however brief, of the rest of the industry. Rather, our aim here is to highlight certain features – or preoccupations, perhaps – that are common to different parts of the industry. Thus, with the exception of molten-salt processes designed to extract the more electropositive elements (notably Al, Mg and Na), and the most electronegative (F_2), all 'inorganic' electrolyses take place in aqueous solution. Given the evident importance of minimizing the cell resistance, the reason is not hard to find: aqueous electrolytes have high conductivities, even at quite moderate temperatures. But as you have seen, an often unavoidable problem then arises, namely the concomitant decomposition of the solvent, water, with H_2 evolution at the cathode, and O_2 evolution at the anode.

A theme running through the industry, then, is the desirability of promoting one or other of these electrode reactions, while suppressing the second. Take the electrolytic extraction of metals such as Cu, Zn and Ni, for example. As you saw in Section 5.3, considerable care is taken to reduce co-production of hydrogen at the cathode to a minimum. But from an energy consumption point of view, equal priority should be given to reducing the overpotential at the other electrode. Thus, if a no more useful product than oxygen can be produced, the anode material should ideally be a *good* catalyst for O_2 evolution (by contrast with chlorine production, for which the reverse situation obtains). Unfortunately, truly effective catalysis of the oxygen electrode reaction remains a largely unrealized goal.

To broaden the discussion a little, we might ask whether it is possible to identify the factors that determine the activity and (in view of the above) the *selectivity* of electrocatalysts. This is a relatively new field of study, and to date only a few general pointers have emerged. Without doubt, platinum remains the best electrocatalyst for a whole range of electrode processes. This can be linked to the importance of the adsorption of intermediate species on the electrode surface: the parallels with heterogeneous catalysis in general are hard to avoid. Of course we are thinking of gaseous products here – it's not much use using a platinum electrode if, after a short time, it becomes coated with another element!

Much the same type of argument can be advanced to explain why the most successful rivals to platinum are invariably based on transition metals, which have unfilled d orbitals available to form bonds with the adsorbate, as discussed in Block 5. To quote Pletcher and Walsh:

> ... it seems that the design of a catalyst requires the placing of transition-metal ions or atoms in a matrix which serves to optimize their electronic configuration and position with respect to each other.
>
> (D. Pletcher and F. Walsh, *Industrial Electrochemistry*, Second edition, 1990, Blackie Academic and Professional, chapter 1)

In practice, the 'matrix' may be an oxide lattice – or it may be created by alloying with another metal, or by forming a metal complex. Certainly, all of these possibilities are the subject of ongoing research.

In service, of course, other factors also come into play. Thus, it is crucial that the catalyst (often a coating on an inert support material, such as DSAs) should be stable and resistant to corrosion under electrolysis conditions, and not easily poisoned. Equally, the *real* surface area of the electrocatalyst is important: the overpotential at a given current load can be reduced by the simple expedient of preparing the electrode with a rough, or otherwise large-area, surface.

The development of effective (and economical) electrocatalysts is recognized to be an important factor in the design of new electrolytic processes. But the priority given to such essentially energy-conserving measures in established industries varies considerably. To cite a familiar example, most metal-extraction plants are sited near sources of cheap hydroelectric power, so present-day technology is designed more for ease of operation than for energy-efficiency: for instance, no serious attention has been given to reducing the overpotential for O_2 evolution at the anodes of such cells. Equally, certain inorganic compounds are prepared electrolytically, but in very low tonnage: there may then be little attempt to optimize the cell design and electrode materials. An extreme example is the anodic oxidation of bromide to bromate (BrO_3^-), for which cells using *solid* platinum electrodes are still in use!

One final point before we turn to 'energy producers' in the next Section: it would be misleading to leave you with the impression that electrolytic processes are concerned exclusively with the preparation of *inorganic* substances. A considerable number of *low-tonnage* organic products are produced by **organic electrosynthesis**.

Nevertheless, it is true to say that this type of production has yet to make a major impact on the chemical industry as a whole. Some of the underlying problems can be seen by taking a brief look at one organic process that has been exploited on a large scale: the cathodic *hydrodimerization* of acrylonitrile (**1**) to adiponitrile (**2**), the latter being an important intermediate in the production of nylon 6–6:

$$\underset{\mathbf{1}}{2CH_2{=}CHCN} + 2H_2O + 2e = \underset{\mathbf{2}}{\begin{array}{l} CH_2CH_2CN \\ | \\ CH_2CH_2CN \end{array}} + 2OH^- \qquad (56)$$

The first point to note is that this process is chemically considerably more complex than the production of chlorine, say (or indeed, any other inorganic compound; see the example in SAQ 8). Fairly obviously, the overall reaction in equation 56 must take place via a series of steps, the first of which is known to be electron transfer to form the *radical anion* **3**. To form adiponitrile, dimerization and protonation must then occur, so (without giving details of the mechanism):

$$CH_2{=}CHCN \xrightarrow{e} \underset{\mathbf{3}}{[CH_2{=}CHCN]^{\overline{\bullet}}} \xrightarrow[2H^+]{\text{dimerization}} \underset{\mathbf{2}}{(CH_2CH_2CN)_2} \qquad (57)$$

■ Suppose, however, that **3** is rapidly protonated, or equally that dimerization is slow. Can you suggest an alternative product that could be formed?

■ The simplest alternative is propanenitrile (propionitrile), **4**, formed as follows:

$$\underset{\mathbf{3}}{[CH_2{=}CHCN]^{\overline{\bullet}}} \xrightarrow[e\ +\ H^+]{H^+} \underset{\mathbf{4}}{CH_3CH_2CN} \qquad (58)$$

The problem, then, is that electron transfer to an organic compound generally produces an intermediate species that has a variety of chemical reactions open to it. It is necessary somehow to direct it along the desired path, and usually this can be achieved only by very precise control of the chemical environment close to the electrode surface: no mean feat! In this case, the trick is to add *quaternary ammonium* salts (that is, with a cation R_4N^+, where R is an alkyl group) to the electrolyte; apparently these adsorb on the surface of the cathode, and render the layer close to it relatively 'aprotic' (proton free).

In terms of energy consumption, can you suggest a second general problem with organic electrosynthesis?

A fundamental difficulty is the combination of *organic* chemistry and *electro*chemistry. The former is usually best suited to an organic solvent, whereas, to minimize the cell resistance, the latter requires a medium of high conductivity (preferably water with a high concentration of electrolyte). This means that the combination invariably leads to some sort of, more or less unsatisfactory, compromise.

Detailed chemistry apart, this remains one of the most serious obstacles to the exploitation of organic electrosynthesis on a high-tonnage industrial scale. However, applications to small scale production have increased dramatically over the past few years, partly as a result of the availability of more appropriate electrochemical cells, and partly through the industry gaining experience in the selection of appropriate electrode materials and membranes. Also, for smaller scale production, processes of less than optimum energy efficiency can be tolerated.

STUDY COMMENT You should now attempt SAQ 8 which, as well as providing further revision of Block 7, links back to information contained in Section 7.2 of this Block.

SAQ 8 Like several other powerful oxidizing agents, potassium permanganate ($KMnO_4$) is prepared electrolytically, the crucial step being the anodic oxidation of aqueous manganate ion (MnO_4^{2-}) in a strongly alkaline solution (pH ≈ 14). Hydrogen evolution takes place at the cathode, so the overall cell reaction is:

$$MnO_4^{2-}(aq) + H_2O(l) = MnO_4^-(aq) + \tfrac{1}{2}H_2(g) + OH^-(aq) \qquad (59)$$

One type of cell uses an electrolyte feed containing MnO_4^- and MnO_4^{2-} at concentrations of 0.20 mol dm^{-3} and 0.25 mol dm^{-3}, respectively. Use information from the S342 *Data Book* to determine the minimum potential for the desired electrolysis at 298.15 K. State any assumptions that you make. What other reactions are possible at this potential, and what material would you recommend for the cathode?

8 BATTERIES AND OTHER ELECTROCHEMICAL POWER SOURCES

We close this Block by taking a brief look at a second major concern of the electrochemical industry – the design of practical 'self-driving' cells. As we suggested in Block 7, provided a suitable cell can be devised, any spontaneous reaction can in theory be 'harnessed' to produce electricity *directly*. You may recall from the Second Level Inorganic Course that this direct conversion of chemical energy into electricity has an inherent advantage over the *indirect* route (involving thermal devices such as boilers and turbines) presently used for most large-scale power generation. The types of cell proposed to exploit this advantage are known collectively as fuel cells: the essential features are outlined in Section 8.1.

Before that, however, we concentrate on their more familiar and more widely used cousins – the **batteries**. Everyday examples include the lead/acid battery in cars, and the wide variety of batteries designed for portable electrical and electronic equipment – ranging from the miniature 'button' cells (used in watches, calculators and the like) to the more substantial versions intended for radios, say, or 'cordless' power tools. Less familiar, perhaps, are the more massive systems at the other end of the range, including both traction batteries – to power electric vehicles (milk floats, for instance) – and stationary batteries – for emergency power supplies, etc.

Many of these batteries are based on the same, relatively few, electrochemical systems – in the sense that the underlying cell reaction is the same; but the examples cited above serve to illustrate the range of performance required. For instance, car batteries remain the most important market for the lead/acid system, but it is also widely used for traction – in vehicles ranging in size from a milk float to a non-nuclear submarine (for propulsion when submerged) – and for back-up (e.g. property security systems) or emergency (e.g. hospital) power supplies. More recently, small cylindrical lead/acid cells suitable for portable equipment have also become available. Overall, more than 10^8 lead/acid batteries are manufactured each year, accounting for some 30% of the world output of lead.

As these remarks suggest, a battery is always designed with a particular end-use in mind, and this will largely determine the characteristics it must have. In general, these will include both the performance of the battery as a source (and store, as you will see in a moment) of electricity, and a variety of other factors – such as overall mass and/or volume, cost, reliability, shelf-life, and so on. Space does not permit us to go into the technical details here. Rather we concentrate on just one aspect: the in-service performance of any battery system revolves around its cell potential, and the way this varies as current is drawn. Our general expression for V again applies:

$$V = (E_{ca} - E_{an}) - |\eta_{ca}| - |\eta_{an}| - IR_{cell} \qquad (46)$$

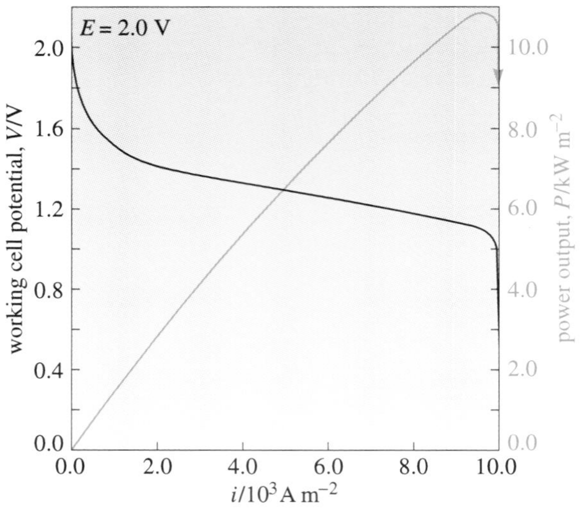

Figure 39 Cell potential (black line) and power output (blue line) versus current density for a hypothetical self-driving cell with the following characteristics: $E = 2.0$ V; $i_{e,\,ca} = i_{e,\,an} = 10$ A m^{-2}; $i_{L,\,ca} = i_{L,\,an} = 1.0 \times 10^4$ A m^{-2}; $\alpha_{ox,\,an} = \alpha_{red,\,ca} = 0.5$; $R_{cell} = 0.1\ \Omega$; electrode areas $A = 1$ cm^2; temperature $T = 300$ K.

only this time, of course, the emf $(E_{ca} - E_{an}) > 0$, so the remaining terms tend to *lower* the cell potential. As with electrolytic cells, the extent to which they do so depends in general on the kinetics of the electrode reactions, the cell resistance and possible concentration polarization effects. We discussed some typical plots of V against i for a self-driving cell in Section 6.1. Figure 27 is repeated here as Figure 39, but we have now included additional information on the power output of the cell.

- Concentrate on the V-curve (black line) in Figure 39. How do you explain the sudden dramatic fall in the cell potential at high current density?

- As the current density approaches the limiting value at one electrode (or, in this case, both electrodes), concentration polarization becomes increasingly important and eventually quenches the cell potential altogether.

The blue curve in Figure 39 shows that the **power output** (defined as $P = IV$) of the cell – the rate at which it supplies electricity – also suffers a catastrophic collapse under these conditions. Thus, the need to avoid concentration polarization is, if anything, more acute for a battery than for an electrolytic cell. But the way around this difficulty must now be rather different, because stirring or agitating the solutions is not normally a viable option for a battery!

The underlying requirement is to ensure a ready supply of the electroactive species at the sites of electron transfer. In most batteries this is achieved by using solid reactants: in many cases the products of the electrode processes are also solid. As indicated in the sketch in Figure 40, in many commercial batteries the electroactive material is formed into a porous paste, which is spread onto a metal grid or plate acting as a current collector.

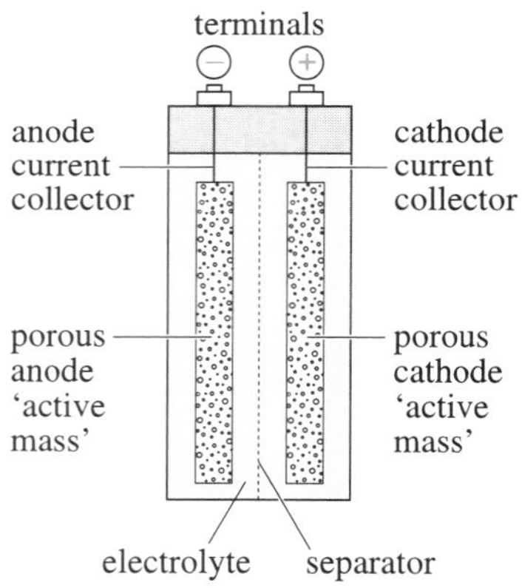

Figure 40 General arrangement of a storage battery.

Here, we shall look, albeit briefly, at the lead/acid car battery, known technically as the *starting, lighting and ignition (SLI) battery*. It is fairly typical of **rechargeable** or **secondary** batteries, so-called because the spontaneous cell reaction can be reversed by passing current through the cell in the opposite direction. As you may recall from Block 7, during *discharge* – producing current, that is – the half-reactions at the two electrodes can be represented as follows:

anode: $$Pb(s) + SO_4^{2-}(aq) = PbSO_4(s) + 2e \quad (60)$$

cathode: $$PbO_2(s) + SO_4^{2-}(aq) + 4H^+(aq) + 2e = PbSO_4(s) + 2H_2O(l) \quad (61)$$

overall: $$Pb(s) + PbO_2(s) + 4H^+(aq) + 2SO_4^{2-}(aq) = 2PbSO_4(s) + 2H_2O(l) \quad (62)$$

with (under standard conditions at 298.15 K)

$$E_{cell} = E_{ca} - E_{an}$$

$$= \{+1.69 - (-0.36)\}\ \text{V} = 2.05\ \text{V}$$

According to the Nernst equation, the actual emf will, of course, depend on the concentration of the sulfuric acid electrolyte; again, you worked through a sample calculation in Block 7. Furthermore, this will change during discharge, as sulfuric acid is consumed and water is formed (equation 62). In practice, the electrolyte composition typically varies from around 40% by weight of H_2SO_4 at full charge (E = 2.15 V) to about 16% when fully discharged (E = 1.98 V). The accompanying change in the density of the electrolyte allows ready monitoring of the state of charge.

As you might expect, the processes actually taking place at the electrodes are more complex than the equations above suggest. Nevertheless, the exchange current densities are high. This, together with an optimum cell design to reduce 'ohmic' and other losses, is why the working potential is rated at around 2 V, and why the cell can supply large currents (and high power outputs) on demand: engine starting, for example, requires a pulse of some 400–450 A for as long as 30 seconds without the potential dropping below 1.2 V. Some typical data are shown in Figure 41.

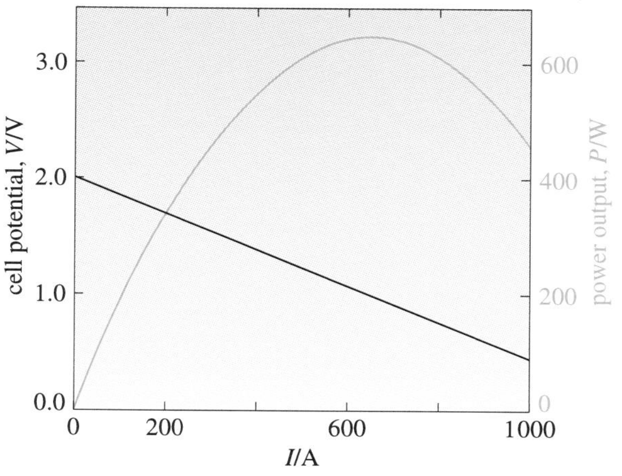

Figure 41 Cell potential (black line) and power output (blue line) as a function of current drawn for an SLI lead/acid battery.

An SLI battery (with three or six lead/acid cells in series) *stores* electricity in the sense that the reactants for the spontaneous processes above are formed in situ – within the cell, that is – when a current is passed from an outside source.* Thus, the battery begins its life as a series of current collectors – usually open grids made from an alloy of lead – each of which is coated with a porous paste containing a mixture of PbO and $PbSO_4$. The pasted grids and the separators (generally microporous polyethene or resin-impregnated paper) are assembled in the battery container, which is then filled with sulfuric acid and sealed. During the initial charging, a low current is 'driven' through the cell, such that the electrode processes are essentially the reverse of those given above. Thus Pb^{2+} is reduced to lead at one set of electrodes (effectively the reverse of equation 60); additives in the paste help to ensure that the lead forms with an open, spongy texture, and hence a large surface area. Meanwhile, Pb^{2+} is oxidized to PbO_2 at the other electrodes (overall, this is roughly the reverse of equation 61). *Recharging*, during the lifetime of the battery, follows much the same pattern, although in this case there will be no PbO in the electrode paste.

SAQ 9 Recent years have seen a trend toward maintenance-free sealed SLI batteries. Can you suggest a danger that might arise if such a battery is overcharged? How do you think this is overcome in practice?

* This contrasts with a simple or *primary* cell, where the 'fuels' are built in during manufacture: such a cell is not rechargeable. Although the familiar range of 'portable' batteries are often primary cells, several less-wasteful secondary cells are available. Fuel cells are different again: here the reactants are stored outside the cell, rather than forming an integral part of its construction (Section 8.1).

With continuing improvements in design and performance, the lead/acid system has held its place in the car-battery market virtually unchallenged. Other battery systems are, of course, in common use, and many more are under active development. Indeed, since the early 1970s there has been a major research and development effort in this area, stemming from the revolution in the electronics industry (with its demand for ever-smaller power supplies) at one extreme, and from the renewed interest in electric vehicles, at the other. Some of the more desirable features of a storage battery are summarized in the answer to SAQ 10.

STUDY COMMENT SAQ 10 helps summarize much of the information presented so far on electrochemical systems; make sure you check it through at some stage.

SAQ 10 Assuming that the object is to maximize the power output of a storage battery, list the features that you would aim to incorporate in a portable form of such a device.

8.1 Fuel cells

Fuel cells were originally conceived as a way of 'burning' primary hydrocarbon fuels (like natural gas, for example) in an electrochemical cell, so as to convert the *full* free energy of oxidation *directly* into electricity. For instance, with methane as fuel (and assuming an acid electrolyte) the desired electrode processes can be represented as follows:

anode: $CH_4(g) + 2H_2O(l) = CO_2(g) + 8H^+(aq) + 8e$ (63)

cathode: $2O_2(g) + 8H^+(aq) + 8e = 4H_2O(l)$ (64)

overall: $CH_4(g) + 2O_2(g) = CO_2(g) + 2H_2O(l);\ \Delta G_m^\ominus(298.15\,K) = -817.9\,kJ\,mol^{-1}$ (65)

- What is the standard emf of the reaction in equation 65 at 298.15 K?
- As written, the half-reactions in equations 63 and 64 involve the transfer of eight electrons, so $E^\ominus = -\Delta G_m^\ominus/8F = 1.06$ V.

But like the more familiar cells discussed above, an acceptable energy efficiency in practice requires that the working potential be close to this theoretical maximum – even for quite high current densities (certainly $>10^3$ A m^{-2}, for a fuel cell to be useful). Unfortunately, there are problems with both electrode reactions. The need to develop better catalysts for the oxygen electrode reaction (equation 64) was mentioned earlier. In view of our discussion of organic electrosynthesis in Section 7.3, it should come as no surprise to learn that the anode reaction (equation 63) presents yet more serious problems. Indeed, it seems doubtful whether conditions could be found that would promote *complete* electrochemical oxidation of a fuel like methane: even if this were possible, it is certain that the catalytic activity of any (economically acceptable) electrode material would be low, hence restricting the current densities obtainable from the cell. However, research continues, especially into the viability of a methanol fuel cell.

Because of these difficulties, some fuel-cell research has taken a different, more indirect, approach. Here, it is accepted that the primary fuel (be it natural gas or another hydrocarbon – or, indeed, even coal in the future) must first be 'reformed' to hydrogen or carbon monoxide, or a mixture of both, *before* being fed to the anode of the fuel cell. The actual cell reaction is then one or other of the following much simpler processes (or a combination of the two):

$$H_2(g) + \tfrac{1}{2}O_2(g) = H_2O(l);\ \Delta G_m^\ominus(298.15\,K) = -237.1\,kJ\,mol^{-1} \quad (66)$$

or

$$CO(g) + \tfrac{1}{2}O_2(g) = CO_2(g);\ \Delta G_m^\ominus(298.15\,K) = -257.2\,kJ\,mol^{-1} \quad (67)$$

In this context, it is worth recalling that effective catalysts for the hydrogen electrode reaction are known, as outlined in Section 7.2.1.

The technology of several such simplified systems is already well advanced. Postulated applications include a 'hybrid' car traction unit (combined with a battery to provide peak power for starting and acceleration) and power generation (both on a large-scale centralized basis, and for smaller-scale supplies at remote sites). Certainly, the hydrogen/oxygen reaction may have a role to play in the exploitation of alternative energy sources. Thus, for example, solar-generated electricity could be used to electrolyse water, the resulting hydrogen being stored and transported to distant sites: fuel cells could then be used to 'reconvert' the hydrogen into electricity, via reaction 66.

The first successful application of a hydrogen/oxygen fuel cell was as an integral part of the US space programme (where money was no object!): an example forms the basis of SAQ 11.

There is now a range of different fuel cells for different types of application. Some run at very high temperatures and, therefore, tend to be used for continuous operation as in power stations. The total world-wide generating capacity of fuel cells at the time of writing (1995) stands at some 40 MW, most of this being in Japan where the largest cell produces an output of 11 MW and serves some 4 000 homes. Other fuel cells run at ambient temperatures, and the power can be switched on and off quickly and frequently; such devices are ideal for powering various forms of transport. Indeed, several buses powered by fuel cells are already in operation in the USA. Vehicles powered in this way emit negligible amounts of exhaust pollutants, are noise-free and are very easy to maintain. However, power generated by fuel cells is still not cheap, and considerable research effort continues into reducing these costs. There is no doubt that fuel cell use will expand dramatically over the coming years.

SAQ 11 Figure 42 shows a sketch of a single H_2/O_2 fuel cell of the type developed for the Apollo space programme: it operated at around 200 °C. State the likely electrode reactions involved and indicate *briefly* how the design features reflect the discussion of practical electrochemical systems in Sections 7 and 8 of this Block.

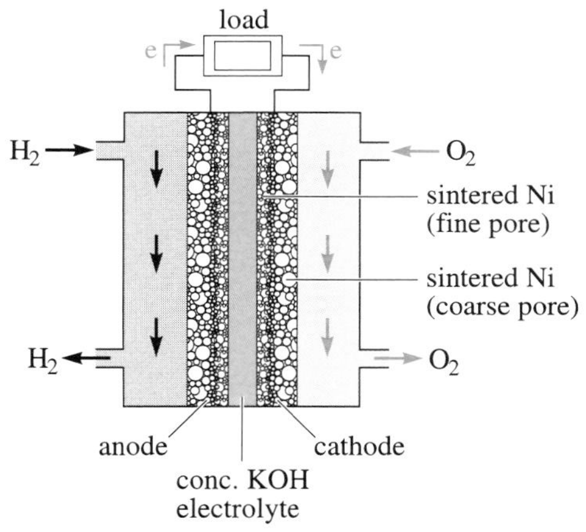

Figure 42 Schematic representation of a single Apollo hydrogen/oxygen fuel cell.

Appendix 1

The derivation of equation 38 in Section 6.1 starts with the expression for the cell potential V, in equation 36:

$$V = {}^{\mathrm{Pt}}\Delta^{\mathrm{Ag}}\phi + {}^{\mathrm{Ag}}\Delta^{\mathrm{Ag^+}}\phi + {}^{\mathrm{Ag^+}}\Delta^{\mathrm{Zn^{2+}}}\phi + {}^{\mathrm{Zn^{2+}}}\Delta^{\mathrm{Zn}}\phi + {}^{\mathrm{Zn}}\Delta^{\mathrm{Pt}}\phi \qquad (36)$$

Recalling that the potential drop across the bulk of the solutions (${}^{\mathrm{Ag^+}}\Delta^{\mathrm{Zn^{2+}}}\phi$) is given by $\Delta\phi_{\mathrm{S}}$ (equation 37), then

$$V = {}^{\mathrm{Pt}}\Delta^{\mathrm{Ag}}\phi + {}^{\mathrm{Ag}}\Delta^{\mathrm{Ag^+}}\phi + {}^{\mathrm{Zn^{2+}}}\Delta^{\mathrm{Zn}}\phi + {}^{\mathrm{Zn}}\Delta^{\mathrm{Pt}}\phi + \Delta\phi_{\mathrm{S}} \qquad (68)$$

When no current is drawn from the cell, $\Delta\phi_{\mathrm{S}} = 0$, $V = E$, and the potential differences have their equilibrium values, so

$$E = {}^{\mathrm{Pt}}\Delta^{\mathrm{Ag}}\phi_{\mathrm{e}} + {}^{\mathrm{Ag}}\Delta^{\mathrm{Ag^+}}\phi_{\mathrm{e}} + {}^{\mathrm{Zn^{2+}}}\Delta^{\mathrm{Zn}}\phi_{\mathrm{e}} + {}^{\mathrm{Zn}}\Delta^{\mathrm{Pt}}\phi_{\mathrm{e}} \qquad (69)$$

Subtracting equation 69 from equation 68 gives the following expression:

$$V - E = ({}^{\mathrm{Pt}}\Delta^{\mathrm{Ag}}\phi - {}^{\mathrm{Pt}}\Delta^{\mathrm{Ag}}\phi_{\mathrm{e}}) + ({}^{\mathrm{Ag}}\Delta^{\mathrm{Ag^+}}\phi - {}^{\mathrm{Ag}}\Delta^{\mathrm{Ag^+}}\phi_{\mathrm{e}}) + ({}^{\mathrm{Zn^{2+}}}\Delta^{\mathrm{Zn}}\phi - {}^{\mathrm{Zn^{2+}}}\Delta^{\mathrm{Zn}}\phi_{\mathrm{e}}) +$$
$$({}^{\mathrm{Zn}}\Delta^{\mathrm{Pt}}\phi - {}^{\mathrm{Zn}}\Delta^{\mathrm{Pt}}\phi_{\mathrm{e}}) + \Delta\phi_{\mathrm{S}} \qquad (70)$$

But, as we pointed out earlier (in Section 4), the potential difference across a metal/metal interface can be taken to be effectively *independent* of the current flow. Thus, the first and fourth terms in equation 70 can be set to zero, which leaves

$$V - E = ({}^{\mathrm{Ag}}\Delta^{\mathrm{Ag^+}}\phi - {}^{\mathrm{Ag}}\Delta^{\mathrm{Ag^+}}\phi_{\mathrm{e}}) + ({}^{\mathrm{Zn^{2+}}}\Delta^{\mathrm{Zn}}\phi - {}^{\mathrm{Zn^{2+}}}\Delta^{\mathrm{Zn}}\phi_{\mathrm{e}}) + \Delta\phi_{\mathrm{S}} \qquad (71)$$

or

$$V - E = ({}^{\mathrm{Ag}}\Delta^{\mathrm{Ag^+}}\phi - {}^{\mathrm{Ag}}\Delta^{\mathrm{Ag^+}}\phi_{\mathrm{e}}) - ({}^{\mathrm{Zn}}\Delta^{\mathrm{Zn^{2+}}}\phi - {}^{\mathrm{Zn}}\Delta^{\mathrm{Zn^{2+}}}\phi_{\mathrm{e}}) + \Delta\phi_{\mathrm{S}} \qquad (72)$$

But the terms in brackets are just the overpotentials at the silver and zinc electrodes, respectively; so equation 72 becomes

$$V - E = \eta(\mathrm{Ag}) - \eta(\mathrm{Zn}) + \Delta\phi_{\mathrm{S}} \qquad (73)$$

Since

$$E = E_{\mathrm{RHE}} - E_{\mathrm{LHE}} = E(\mathrm{Ag^+|Ag}) - E(\mathrm{Zn^{2+}|Zn})$$

then

$$V = \{E(\mathrm{Ag^+|Ag}) - E(\mathrm{Zn^{2+}|Zn})\} + \eta(\mathrm{Ag}) - \eta(\mathrm{Zn}) + \Delta\phi_{\mathrm{S}}$$

which is equation 38 in the text.

OBJECTIVES FOR BLOCK 8

Now that you have completed Block 8, you should be able to do the following things:

1 Recognize valid definitions of and use in a correct context the terms, concepts and principles printed in bold in the text and collected in the following Table.

List of scientific terms, concepts and principles used in Block 8

Term	Page No.
batteries	51
Butler–Volmer equation	17
cell potential–current density relations for complete cells	37
cell separators	41
charge separation	6
chlor-alkali industry	40
concentration polarization	36
conductivity, κ	35
current density, i	12
current efficiency	28
driven cell	31
electrical double layer	6
electrical resistance (of solution), R_S	32
electrocatalysis	24
electrosynthesis – inorganic and organic	48
equilibrium potential difference, $\Delta\phi_e$	7
exchange current density, i_e	17
fuel cell	54
limiting current density, i_L	36
minimizing cell resistance	35
net current density, $i_{net} = i_{ox} - i_{red}$	15
Ohm's law	32
outer Helmholtz plane, OHP	11
overpotential, η	15
overpotential for net oxidation (at an anode) and net reduction (at a cathode), η_{an} and η_{ca}	34
potential difference, $\Delta\phi$	6
power output, P	52
secondary (rechargeable) battery	52
self-driving cell	32
Tafel plot	12
transfer coefficient, α	17
working cell potential, V	32

2 Describe how a potential difference arises at an interface, and discuss in what ways the equilibrium potential difference differs from the electrode potential value. (SAQ 2)

3 Given the composition of an electrochemical cell, write down an expression for the measured potential difference, V, in terms of the potential differences across all the interfaces involved. (SAQ 1)

4 Outline the features of the Helmholtz model of the electrified interface.

5 Sketch graphs of reaction rate versus time, net reaction rate versus time, net reaction rate versus potential difference, and current density versus overpotential for various electrochemical processes. (Exercise 1)

6 Deduce the Tafel equation from the Butler–Volmer equation at both positive and negative values of the overpotential. (SAQ 6)

7 Describe briefly how Tafel data are obtained experimentally and how such data are presented.

8 Construct Tafel plots from given experimental data and so determine values of i_e and α. (SAQ 3)

9 Given appropriate Tafel data, speculate as to the probable reaction mechanism, by calculating α values from the relationships given in Section 5.2. (SAQ 4)

10 Given various experimental or hypothetical data, predict or rationalize the likely reaction at an electrode under particular conditions. (SAQs 5 and 8; Exercise 3)

11 Write an expression for the overall potential, V, of a cell in terms of the cell emf E, the overpotentials η_{an} and η_{ca} at the anode and cathode, respectively, and the internal cell resistance R_{cell} – and use this expression to explain why: (a) $|V|$ is always smaller than $|E|$ for a self-driving cell; but (b) larger than $|E|$ for a driven cell.

12 State the factors that determine the resistance of a solution and the limiting current density at an electrode, and indicate how these can be minimized in industrial cells. (SAQs 10 and 11)

13 Given appropriate information, select the most suitable electrode material for a specified electrode reaction, or rationalize the choice of material used in practice. (SAQs 8, 9 and 11; Exercise 3)

14 Using the ideas outlined in Objectives 10–13, predict or rationalize the potential required for a given electrolytic process in terms of the factors that determine:

(a) the cell emf;

(b) the overpotential at each electrode;

(c) the current efficiency; and

(d) the internal cell resistance.

(SAQs 7, 8 and 9; Exercises 2 and 3)

15 Predict or rationalize the power output of a given storage battery or fuel cell in terms of the factors outlined in Objective 14. (SAQ 10)

16 Rationalize the design of a given battery or fuel cell in terms of the factors outlined in Objective 14. (SAQ 11)

SAQ ANSWERS AND COMMENTS

SAQ 1 (Objective 3)

Reading from the right, the potential difference across the electrochemical cell will be given by the expression:

$$V = {}^{\mathrm{Pt}}\Delta^{\mathrm{Zn}}\phi + {}^{\mathrm{Zn}}\Delta^{\mathrm{Zn^{2+}}}\phi + {}^{\mathrm{Zn^{2+}}}\Delta^{\mathrm{Cu^{2+}}}\phi + {}^{\mathrm{Cu^{2+}}}\Delta^{\mathrm{Cu}}\phi + {}^{\mathrm{Cu}}\Delta^{\mathrm{Pt}}\phi$$

The emf of the cell is determined under equilibrium conditions, so

$$V_e = E = {}^{\mathrm{Pt}}\Delta^{\mathrm{Zn}}\phi_e + {}^{\mathrm{Zn}}\Delta^{\mathrm{Zn^{2+}}}\phi_e + {}^{\mathrm{Zn^{2+}}}\Delta^{\mathrm{Cu^{2+}}}\phi_e + {}^{\mathrm{Cu^{2+}}}\Delta^{\mathrm{Cu}}\phi_e + {}^{\mathrm{Cu}}\Delta^{\mathrm{Pt}}\phi_e$$

Assuming liquid-junction potentials have been eliminated,

$$E = {}^{\mathrm{Pt}}\Delta^{\mathrm{Zn}}\phi_e + {}^{\mathrm{Zn}}\Delta^{\mathrm{Zn^{2+}}}\phi_e + {}^{\mathrm{Cu^{2+}}}\Delta^{\mathrm{Cu}}\phi_e + {}^{\mathrm{Cu}}\Delta^{\mathrm{Pt}}\phi_e$$

Platinum wires are used, so the electrode potential of the (Cu^{2+}|Cu) system will be given by (cf. equation 8 in Section 2.2):

$$E(\mathrm{Cu^{2+}|Cu}) = {}^{\mathrm{Pt}}\Delta^{\mathrm{Cu}}\phi_e + {}^{\mathrm{Cu}}\Delta^{\mathrm{Cu^{2+}}}\phi_e$$

and the electrode potential of the (Zn^{2+}|Zn) system will be given by:

$$E(\mathrm{Zn^{2+}|Zn}) = {}^{\mathrm{Pt}}\Delta^{\mathrm{Zn}}\phi_e + {}^{\mathrm{Zn}}\Delta^{\mathrm{Zn^{2+}}}\phi_e$$

Thus the expression for E derived above can be written:

$$E = \{{}^{\mathrm{Pt}}\Delta^{\mathrm{Zn}}\phi_e + {}^{\mathrm{Zn}}\Delta^{\mathrm{Zn^{2+}}}\phi_e\} - \{({}^{\mathrm{Pt}}\Delta^{\mathrm{Cu}}\phi_e + {}^{\mathrm{Cu}}\Delta^{\mathrm{Cu^{2+}}}\phi_e)\}$$
$$= E(\mathrm{Zn^{2+}|Zn}) - E(\mathrm{Cu^{2+}|Cu})$$

Notice that this result is also in accord with the conventions established in Block 7, where the emf of a cell was written as $E_{\mathrm{cell}} = E_{\mathrm{RHE}} - E_{\mathrm{LHE}}$, E_{RHE} and E_{LHE} being the electrode potentials of the half-reactions at the right-hand electrode and left-hand electrode, respectively.

SAQ 2 (Objective 2)

(a) False. At the instant of immersion of the electrode, the *potential difference* ($\Delta\phi$) is zero: the potential difference then changes to its equilibrium value. The value of the overpotential at $t = 0$ will have its highest value for this system.

(b) False. At equilibrium, by convention the *electrode potential* is defined as being 0.0 V. In addition, by convention, we set the sum of ${}^{\mathrm{Pt,\,H_2}}\Delta^{\mathrm{H^+}}\phi_e + {}^{\mathrm{Pt}}\Delta^{\mathrm{Pt,\,H_2}}\phi_e$ as 0.0 V. This doesn't necessarily mean that the *equilibrium potential difference will* be 0.0 V.

(c) True. By definition, if the system is at equilibrium the overpotential must be zero.

(d) False. When the overpotential is negative, *net* reduction is occurring. Thus, oxidation as well as reduction will be taking place under these conditions. If the overpotential is *large and negative* the rate of the oxidation reaction could be very low.

(e) False. When the overpotential is 0.0 V, the system is at equilibrium. Under these conditions the rate of the reduction reaction equals the rate of the reverse, oxidation, reaction, but the rates are unlikely to be zero.

SAQ 3 (Objective 8)

(a) Because the data are obtained using positive values of the overpotential the *net* reaction must be oxidation, that is:

$$Fe^{2+}(aq) = Fe^{3+}(aq) + e$$

(b) The current density is determined from the current by dividing the values by the area of the electrode. Thus, the following table can be constructed.

η/mV	+50	+100	+150	+200	+250
i/mA cm^{-2}	4.4	12.5	29.0	65.5	149
i/A m^{-2}	44	125	290	655	1 490
log | i/A m^{-2} |	1.64	2.10	2.46	2.82	3.17

The Tafel plot constructed from these data is shown in Figure 43. The slope of the graph is 140 mV and the intercept on the log | i | axis (at $\eta = 0$) = 1.39.

From equation 22,

$$\text{slope} = \frac{2.303RT}{\alpha_{ox}F}$$

so

$$\alpha_{ox} = (2.303RT)/(\text{slope} \times F)$$

With $T = 298.15$ K (25 °C) and $F = 96\,485$ C mol^{-1}

$$\alpha_{ox} = \frac{2.303 \times (8.314 \text{ J K}^{-1} \text{ mol}^{-1}) \times (298.15 \text{ K})}{(140 \times 10^{-3} \text{ V}) \times (96\ 485 \text{ C mol}^{-1})}$$

$$= 0.42 \text{ J V}^{-1} \text{ C}^{-1}$$

$$= 0.42 \text{ (since V = J C}^{-1})$$

When $\eta = 0$, log | i | = log | i_e |. From Figure 43,

$$\text{intercept} = \log |i_e/\text{A m}^{-2}| = 1.39$$

so

$$i_e = 10^{1.39} = 24.5 \text{ A m}^{-2}$$

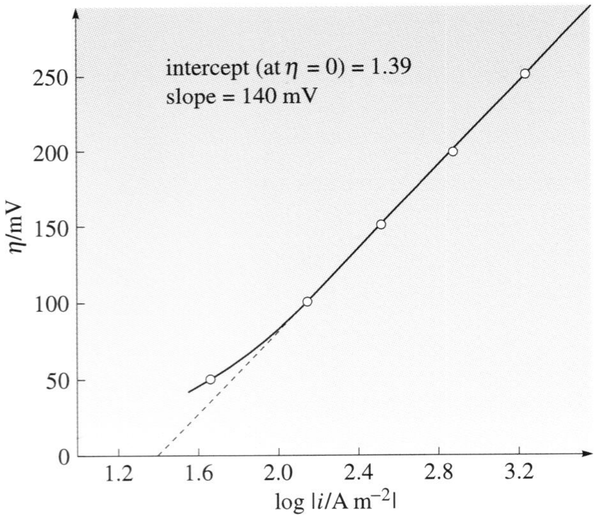

Figure 43 A Tafel plot for the oxidation of $Fe^{2+}(aq)$ to $Fe^{3+}(aq)$.

SAQ 4 (Objective 9)

For the proposed mechanism the following values of α can be calculated from equations 28 and 29:

	Rate-limiting step		
	(i)	(ii)	(iii)
α_{red}	0.0 ($\gamma_B = 0$; $n = 0$)	0.5 ($\gamma_B = 0$; $n = 1$)	1.5 ($\gamma_B = 1$; $n = 1$)
α_{ox}	2.0 ($\gamma_A = 2$; $n = 0$)	1.5 ($\gamma_A = 1$; $n = 1$)	0.5 ($\gamma_A = 0$; $n = 1$)

The experimental Tafel slopes of −120 mV and 40 mV yield α values of $\alpha_{red} = 0.5$ and $\alpha_{ox} = 1.5$ at room temperature. This situation is expected from the proposed mechanism if step (ii) is rate-limiting.

For the other possible mechanism, values of α_{ox} and α_{red} are calculated to be as follows:

	Rate-limiting step				
	(i)	(ii)	(iii)	(iv)	(v)
α_{red}	0 ($\gamma_B = 0$; $n = 0$)	0 ($\gamma_B = 0$; $n = 0$)	0 ($\gamma_B = 0$; $n = 0$)	0.5 ($\gamma_B = 0$; $n = 1$)	1.5 ($\gamma_B = 1$; $n = 1$)
α_{ox}	2.0 ($\gamma_A = 2$; $n = 0$)	2.0 ($\gamma_A = 2$; $n = 0$)	2.0 ($\gamma_A = 2$; $n = 0$)	1.5 ($\gamma_A = 1$; $n = 1$)	0.5 ($\gamma_A = 0$; $n = 1$)

Thus, this mechanism has the same Tafel slopes as the experimental values if step (iv) is rate-limiting. Therefore this mechanism fits the experimental Tafel data just as well as the proposed mechanism. It is other factors that make this particular mechanism unlikely, such as the fact that the first step is termolecular.

SAQ 5 (Objective 10)

From the S342 *Data Book*, $E^{\ominus}(H^+|H_2) = 0.0$ V and $E^{\ominus}(Ni^{2+}|Ni) = -0.24$ V.

Applying the Nernst equation to the $(H^+|H_2)$ couple (see Block 7, SAQ 23):

$$E(H^+|H_2) = (-0.059\,2\text{ V})\,\text{pH at }300\text{ K}$$

so at pH = 4.0,

$$E(H^+|H_2) = -0.24\text{ V}$$

When $a(Ni^{2+}) = 1.0$,

$$E(Ni^{2+}|Ni) = E^{\ominus}(Ni^{2+}|Ni) = -0.24\text{ V}$$

Thus, at any particular imposed value of the potential difference, the overpotentials for the two reactions are essentially identical. The value of i_e for formation of hydrogen on nickel is given as 10^{-1} A m^{-2} (Table 1) under standard concentration conditions (i.e. at pH = 0). Increasing the pH to 4.0 *reduces* [H^+] by a factor of 10^4, and so decreases this value by a factor of 10^2, yielding an i_e value of 10^{-3} A m^{-2}. This value is considerably higher than the value for formation of nickel on nickel (which will be 10^{-5} A m^{-2} as in Table 1, assuming a solution with an activity of nickel ions of 1.0 has a concentration of nickel ions of 1 mol dm^{-3}), so that if the Tafel plots have the same slope hydrogen gas will always be produced much faster than nickel metal. Since nickel metal *is* produced successfully under these conditions, the Tafel slope for nickel production must be *less negative* than the Tafel slope for hydrogen production, such that the two Tafel plots cross. In this way a particular potential difference can be applied at which the value of i(Ni) > i(H). The situation is shown schematically in Figure 44.

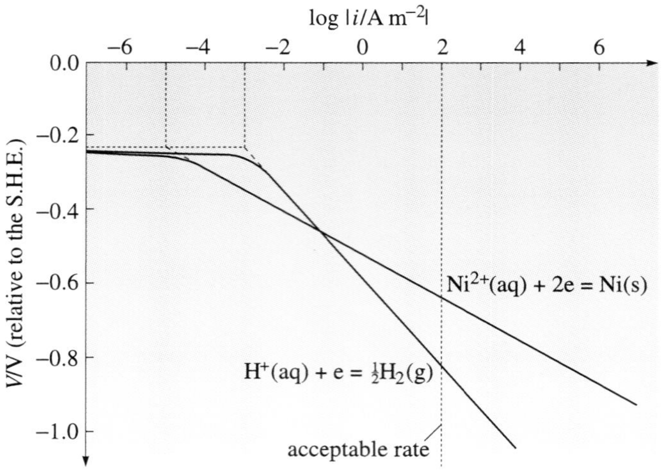

Figure 44 Schematic plots of V (relative to the S.H.E.) versus $\log|i|$ for the competing processes of nickel deposition and hydrogen liberation at a nickel electrode (with different Tafel slopes).

SAQ 6 (Objective 6)

The first term will dominate when the overpotential is large and positive, which implies that there will be *net oxidation* at the electrode. Under these circumstances, equation 14 can be applied to the reaction at the anode to give (compare equation 19, Section 3.3):

$$i = i_{\text{e, an}} \exp(\alpha_{\text{ox, an}} F\eta_{\text{an}}/RT)$$

or

$$i/i_{\text{e, an}} = \exp(\alpha_{\text{ox, an}} F\eta_{\text{an}}/RT)$$

Taking natural logarithms, this becomes

$$\ln(i/i_{\text{e, an}}) = \alpha_{\text{ox, an}} F\eta_{\text{an}}/RT$$

or

$$\eta_{\text{an}} = (2.303RT/\alpha_{\text{ox, an}}F)\log(i/i_{\text{e, an}})$$

SAQ 7 (revision and Objective 14)

(a) Applying the Nernst equation to this couple (notice again that this is written as a reduction) gives:

$$E = E^{\ominus}(\text{Cl}_2|\text{Cl}^-) - \frac{RT}{2F}\ln\frac{\{a(\text{Cl}^-)\}^2}{a(\text{Cl}_2)}$$

$$= E^{\ominus}(\text{Cl}_2|\text{Cl}^-) - \frac{2.303RT}{2F}\log\frac{\{a(\text{Cl}^-)\}^2}{a(\text{Cl}_2)}$$

Assuming ideal behaviour and unit activity for the gas (that is, $p(\text{Cl}_2) = p^{\ominus} = 1$ bar), neglecting the activity coefficient of Cl^- (that is, *assuming* $a(\text{Cl}^-) = c(\text{Cl}^-)/c^{\ominus}$, which is probably a gross approximation for such a concentrated solution), and taking $E^{\ominus}(\text{Cl}_2|\text{Cl}^-) = 1.36$ V:

$$E = (1.36 - 0.059\,2\log 4.0)\text{ V} = 1.32\text{ V}$$

(b) From the answer to Exercise 2, at pH 4:

$$E = E^{\ominus} - (0.059\,2\text{ V})\,\text{pH}$$

$$= (1.23 - 0.237)\text{ V} = 0.99\text{ V}$$

(c) Again, from the answer to Exercise 2, pH 14 $\{a(\text{OH}^-) = 1\}$ is the standard state for this couple so,

$$E = E^{\ominus}(\text{H}_2\text{O}|\text{H}_2,\text{OH}^-) = E(\text{H}^+|\text{H}_2)\text{ at pH 14}$$

$$= -0.83\text{ V}$$

The answers to parts (b) and (c) again assume ideal behaviour and unit activity for the gases and unit activity for the solvent – water.

SAQ 8 (Objectives 10, 13 and 14)

Equation 59 can be 'decomposed' into the following half-reactions:

anode: $MnO_4^{2-}(aq) = MnO_4^-(aq) + e$; $E^{\ominus}_{an} = +0.56$ V

cathode: $H_2O(l) + e = \frac{1}{2}H_2(g) + OH^-(aq)$; $E^{\ominus}_{ca} = -0.83$ V

Then, $E_{cell} = E_{ca} - E_{an}$ where E_{ca} and E_{an} are the *reduction* potentials of the couples above under the conditions specified in the question. In particular E_{an} refers to the following reduction process:

$$MnO_4^-(aq) + e = MnO_4^{2-}(aq)$$

Applying the Nernst equation to this couple gives:

$$E_{an} = E^{\ominus}_{an} - \frac{RT}{F}\ln\frac{a(MnO_4^{2-})}{a(MnO_4^-)}$$

$$= +0.56\text{ V} - \frac{2.303RT}{F}\log\frac{0.25}{0.20}$$

$$= +0.55\text{ V}$$

where the activity coefficients of MnO_4^- and MnO_4^{2-} have been neglected.

With $a(OH^-) = 1$ (at pH 14), E_{ca} has its standard value of –0.83 V, so

$$E_{cell} = (-0.83 - 0.55)\text{ V} = -1.38\text{ V}$$

Thus, the minimum applied potential for electrolysis is 1.38 V.

As in the production of chlorine, the minimum potential applied is more than sufficient to liberate O_2 at the anode. In practice this is a significant problem, resulting in a lowering of the current efficiency, and hence an increase in the energy consumption.

The (strongly alkaline) conditions used for this process suggest that the cathode material developed for the chlor-alkali diaphragm cell (steel coated with nickel alloys of large surface area, Section 7.2.1) could probably be used and would help to reduce the overpotential for hydrogen evolution.

SAQ 9 (Objectives 13 and 14)

To recharge the battery, the cell reaction must be 'driven' in reverse (as outlined in the text): this requires an applied potential greater than the cell emf. As for many industrial electrolyses, this is sufficient (on thermodynamic grounds) to decompose water to H_2 and O_2. That this does not happen *throughout* the recharging process can be ascribed to the low rate, both of the oxygen electrode reaction in general, and of the hydrogen electrode reaction on a Pb-based substrate ($i_e \sim 10^{-8}$ A m^{-2} from Table 1). If current continues to be passed once the cell is fully charged, however, gas evolution will commence, leading (in a sealed unit) to a potentially explosive situation. The problem is solved by incorporating a 'chemical' catalyst that allows controlled recombination of the gases.

SAQ 10 (Objectives 12 and 15)

From the discussion in this and previous Sections, the power output (see Figures 39 and 41) is enhanced by maintaining a high working potential. Thus, one would aim to incorporate the following:

1 A cell reaction with high emf.

2 Electrode reactions having low overpotentials, i.e. high exchange current densities and high α values, if necessary with the help of suitable electrocatalysts (and/or elevated temperature).

3 A physical structure of the electrodes that gives a large effective surface area and reduces any concentration polarization difficulties as far as possible.

4 A minimum value of R_{cell}, by using (a) an electrolyte of high conductivity (a concentrated aqueous solution, for example); (b) a cell design giving the minimum distance possible between the electrodes; and (c) the most efficient low-resistance separator available. Operation at an elevated temperature would also help here.

5 Electroactive materials of low molar mass, combined with the lightest possible materials of construction (compatible with stability and mechanical strength, that is) to decrease the mass of the battery and hence increase its portability.

In practice, many other factors may need to be taken into account, including the often overriding importance of cost. For example, points 1, 2 and 5 above have encouraged the study of high-temperature systems based on reactions of lithium, sodium, or aluminium – with molten electrolytes – where there are major problems associated with corrosion and safety!

SAQ 11 (Objectives 12, 13 and 16)

The reactions of interest are:

anode: $H_2(g) + 2OH^-(aq) = 2H_2O(l) + 2e$

cathode: $\frac{1}{2}O_2(g) + H_2O(l) + 2e = 2OH^-(aq)$

overall: $H_2(g) + \frac{1}{2}O_2(g) = H_2O(l)$

The following features are worthy of note:

1 The electrolyte is *concentrated* KOH, the electrodes are close together, and the cell is operated at a high temperature: all three features reduce the internal resistance of the cell.

2 The electrodes are made of nickel – a good electrocatalyst for the hydrogen electrode reaction (Section 7.2.1). (It is also one of the more effective metals for the oxygen electrode reaction.) Raising the temperature also improves the kinetics, of course.

3 The electrodes have a structure that is porous, and hence of a large surface area, which increases the current for a given current density.

In this case, however, the porous structure also serves another, more fundamental function, namely to allow intimate contact between the gaseous reactant, the liquid electrolyte and the solid electrocatalyst at each electrode. In practice, it is very important to maintain a stable three-phase boundary within the pores: here, this is achieved by means of a double-porous structure, as indicated in Figure 42.

One final point. As we stressed in Section 7.2.2, the kinetics of the oxygen electrode reaction are more rapid in alkaline than in acid media – hence the choice of KOH for the Apollo cell (Figure 42). For terrestrial use, however, atmospheric oxygen (air) is the preferred (that is, cheapest!) cathode reactant, and this has encouraged a shift to acidic electrolytes: with alkaline solutions, CO_2 present in the atmosphere would lead to a build-up of solid carbonate within the pores of the electrodes.

ANSWERS TO EXERCISES

Exercise 1 (Objective 5)

The required sketches are shown in Figures 45–49.

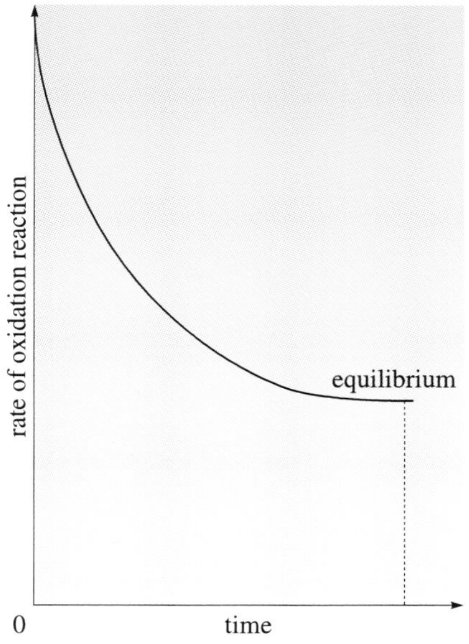

Figure 45 A sketch of the rate of the oxidation reaction versus time.

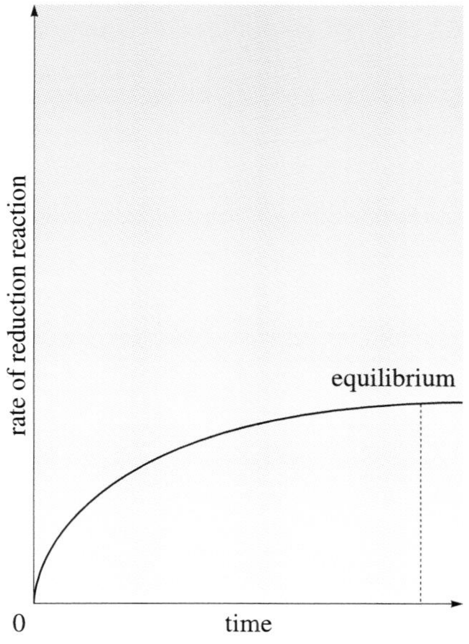

Figure 46 A sketch of the rate of the reduction reaction versus time.

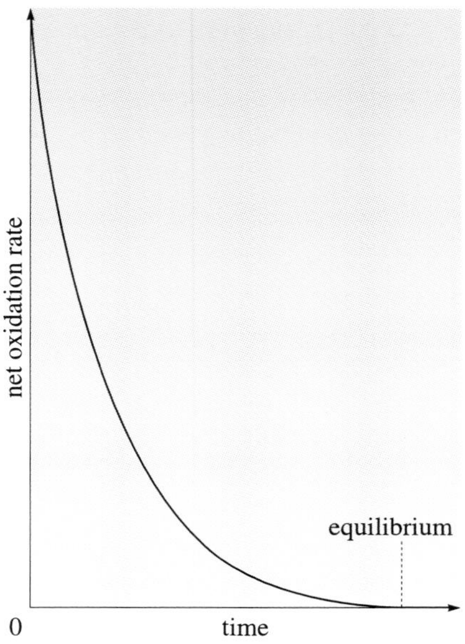

Figure 47 A sketch of the *net* oxidation rate versus time.

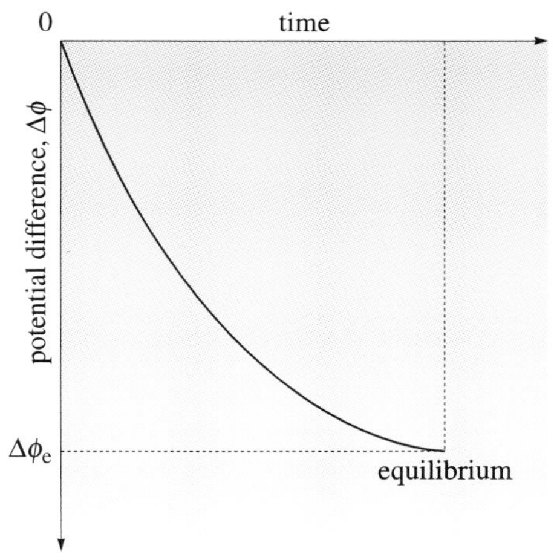

Figure 48 A sketch of the potential difference versus time.

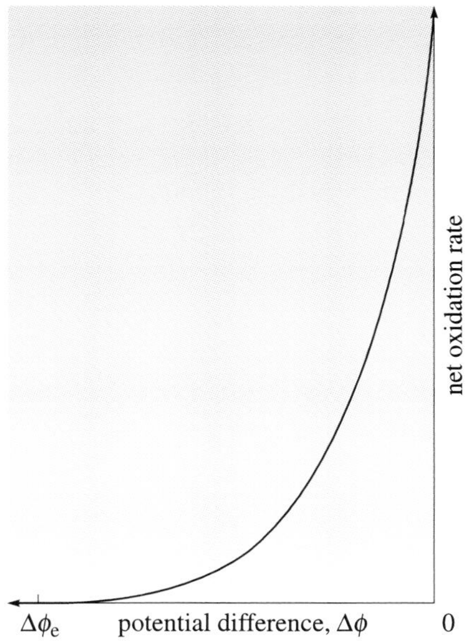

Figure 49 A sketch of the *net* oxidation rate versus potential difference.

(a) If reaction 1 is the most likely reaction, then the rate of this reaction will be highest at the start of the reaction. As the reaction proceeds, the electrode will become negatively charged and the solution positively charged; a potential difference will arise across the interface. This potential difference will have the effect of reducing the rate of reaction 1 because it will become progressively more difficult for zinc ions to enter the solution. Thus a plot of reaction rate versus time is likely to resemble Figure 45.

As reaction 1 slows down, so the rate of the reverse of reaction 1 increases until dynamic equilibrium is reached; at this point the rates of the forward and reverse reactions must be equal. So the rate of reaction 1 will decrease with time – not to zero, but to some constant value.

(b) Similarly the rate of the reverse reaction will increase with time until the same constant value is reached (Figure 46).

(c) The *net* rate of the oxidation reaction at any time is given by the rate of the oxidation reaction *minus* the rate of the reduction reaction (the reverse of reaction 1). Therefore, a plot of the net oxidation rate versus time starts at a large positive value and decreases until equilibrium is reached, at which point the net rate is zero (Figure 47).

(d) At the moment of immersion of the zinc electrode in the solution (time $t = 0$), the potential difference across the interface is zero. As the reaction proceeds, the potential difference becomes more *negative* (since ϕ_M becomes more negative and ϕ_S more positive) until equilibrium is reached; at this point the potential difference across the interface becomes constant and equal to the value of the equilibrium potential difference, $\Delta\phi_e$ (Figure 48).

(e) When the potential difference across the interface is zero, the net oxidation rate is large and positive. When the potential difference has attained the equilibrium value the net rate of the oxidation reaction is zero. Therefore a plot of *net* oxidation rate versus potential difference should resemble Figure 49.

Exercise 2 (Objective 14)

(a) *Anode*: Remember that $E^{\ominus}_{an}$ refers to the couple written as a *reduction* reaction as follows (the reverse of equation 50):

$$\tfrac{1}{2}O_2(g) + 2H^+(aq) + 2e = H_2O(l);\ E^{\ominus}_{an} = 1.23\ V$$

Applying the Nernst equation to this couple, assuming $a(H_2O) = 1$ and $a(O_2) = p(O_2)/p^{\ominus} = 1$ (see Block 7, Section 6.2.3), gives

$$E_{an} = (1.23 - 0.059\,2\,pH)\ V$$

The ion product of water, K_w (= $a(H^+) \times a(OH^-) = 10^{-14}$), links the activities of H^+ and OH^- in any aqueous solution. Thus, when $a(OH^-) = 1.0$, $a(H^+) = 10^{-14}$ and pH = 14, so $E_{an} = 0.40$ V, the value given for the *standard* reduction potential of the couple in equation 48. (Remember that $a(OH^-) = 1$ *is* the standard state for this system.)

Cathode: Similarly, applying the Nernst equation to the ($H^+|H_2$) couple in equation 51 (with $a(H_2) = p(H_2)/p^{\ominus} = 1$) gives (again from Block 7):

$$E_{ca} = (-0.059\,2\,pH)\ V$$
$$= -0.83\ V \text{ at pH} = 14$$

which is the value quoted for $E^{\ominus}_{ca}$ of the couple in equation 49.

Overall: The minimum potential for electrolysis is just the magnitude of the emf of the overall reaction. Both electrode reactions are affected in the same way by pH and so, with a *single* electrolyte (as indicated in Figure 30), the overall emf at 298.15 K must be:

$$E_{cell} = E_{ca} - E_{an}$$

either

$$E_{cell} = E^{\ominus}(51) - E^{\ominus}(50) = 0 - 1.23\ V$$

or

$$E_{cell} = E^{\ominus}(49) - E^{\ominus}(48) = (-0.83 - 0.40)\ V = -1.23\ V$$

Thus, the minimum potential for electrolysis is 1.23 V.

(b) According to the discussion in Block 7 (Section 7.3), the temperature-dependence of the emf is determined by the value of $\Delta S^{\ominus}_m$ for the overall reaction: in this case,

$$\Delta S^{\ominus}_m(298.15\ K) = S^{\ominus}(H_2,\ g) + \tfrac{1}{2}S^{\ominus}(O_2,\ g) - S^{\ominus}(H_2O,\ l)$$
$$= \{130.7 + (\tfrac{1}{2} \times 205.1) - 69.9\}\ J\ K^{-1}\ mol^{-1}$$
$$= 163.4\ J\ K^{-1}\ mol^{-1}$$

Assuming that $\Delta S_m^\ominus$ (and implicitly $\Delta H_m^\ominus$) does not change with temperature:

$$E_2^\ominus - E_1^\ominus = \frac{\Delta S_m^\ominus (T_2 - T_1)}{nF}$$

With T_1 = 298.15 K, T_2 = 353.15 K (80 °C), n = 2 (from equations 48 and 49, or 50 and 51), F = 96 485 C mol^{-1} and $E_1^\ominus$ = −1.23 V, this gives

$$E_2^\ominus(80\,°\mathrm{C}) = \frac{(163.4\ \mathrm{J\,K^{-1}\,mol^{-1}}) \times (55\ \mathrm{K})}{(2 \times 96\ 485\ \mathrm{C\,mol^{-1}})} - 1.23\ \mathrm{V}$$
$$= (+0.047 - 1.23)\ \mathrm{V} = -1.18\ \mathrm{V}$$

Thus, the minimum potential for electrolysis is 1.18 V.

(c) In each case, the *magnitude* of the overpotential is given by equation 47, as:

$$|\eta| = \left(\frac{2.303RT}{\alpha F}\right) \log\left(\frac{|i|}{i_e}\right) \qquad (47)$$

The first term in brackets is just the Tafel slope for the electrode reaction in question (that is, the slope of a plot of η against log $|i|$, as discussed in Section 3.3). Thus, with $i = 1.5 \times 10^3$ A m^{-2}, and taking data from Table 4:

$$\eta_{an} = 0.095\ \mathrm{V}\ \log\left(\frac{1.5 \times 10^3\ \mathrm{A\,m^{-2}}}{4.2 \times 10^{-2}\ \mathrm{A\,m^{-2}}}\right)$$
$$= 0.43\ \mathrm{V}$$

$$|\eta_{ca}| = 0.140\ \mathrm{V}\ \log\left(\frac{1.5 \times 10^3\ \mathrm{A\,m^{-2}}}{2.0\ \mathrm{A\,m^{-2}}}\right)$$
$$= 0.40\ \mathrm{V}$$

So η_{ca} = −0.40 V.

Neglecting (as we must at this stage) the internal resistance of the cell (the final term in equation 46), the cell potential becomes

$$V = (-1.18 - 0.40 - 0.43)\ \mathrm{V} = -2.01\ \mathrm{V}$$

Thus, under these conditions, the applied potential required is estimated to be 2.01 V.

In this case, raising the temperature lowers the magnitude of the minimum cell potential; but this will not always be so because it depends on the sign of $\Delta S_m^\ominus$, as we noted for self-driving cells in Section 6.1. However, recall that the exchange current density, i_e, is a measure of the rate for an electrode reaction: like the rate of any other reaction, it is invariably *increased* by raising the temperature – and increasing i_e *reduces* the overpotential at each electrode, for a given current density i (equation 47).

In practice, the type of electrolyser shown in Figure 30 operates at a potential of around 2.1 V per cell, but the breakdown is usually rather different from that calculated above. In particular, measures we shall discuss in Section 7 can be taken to reduce further the overpotentials at the electrodes. But we also need to take into account the effect of these measures on the cell's resistance.

Exercise 3 (Objectives 10, 13 and 14)

(a) Applying the Nernst equation to the couple at the cathode gives:

$$E_{ca} = E^{\ominus} - \frac{RT}{2F}\ln\left(\frac{a\{Na(Hg)\}}{a(Na^{+})a(Hg)}\right)^2$$

$$= E^{\ominus} - \frac{RT}{F}\ln\left(\frac{a\{Na(Hg)\}}{a(Na^{+})a(Hg)}\right)$$

$$= E^{\ominus} - \frac{2.303RT}{F}\log\left(\frac{a\{Na(Hg)\}}{a(Na^{+})a(Hg)}\right)$$

where $a\{Na(Hg)\}$ represents the activity of sodium in the amalgam. You have no information about this, so can only assume it to be unity: with $a(Hg) = 1$ and setting $\gamma_{\pm}$ for Na^+ to unity, the expression above becomes (at 298.15 K)

$$E_{ca} = (-1.89 + 0.059\,2 \log 4.0)\ \text{V} = -1.85\ \text{V}$$

Taking information from Table 5, $E_{an} = 1.32$ V, so the cell emf (at 298.15 K) is

$$E(\text{overall}) = E_{ca} - E_{an} = (-1.85 - 1.32)\ \text{V} = -3.17\ \text{V}$$

Thus the minimum potential for electrolysis is 3.17 V.

(b) Under these conditions (that is, at pH 4 and $T = 298.15$ K), $E(H^+|H_2) = (-0.059\,2\ \text{V pH}) = -0.24$ V. This is considerably *less* negative than the potential required for discharge of Na^+ ions according to equation 53 (as calculated above), so hydrogen evolution should be the favoured process on thermodynamic grounds. Put in a different way, the minimum potential for electrolysis will be *less* for the production of hydrogen than for the discharge of Na^+ ions. Evidently, a mercury cell functions only because the H_2 evolution reaction is *kinetically* hindered both on a mercury surface and at sodium amalgam.

A hint that this is so comes from the very low value, $i_e \sim 10^{-8}\ \text{A m}^{-2}$ (Table 1), for H_2 evolution on mercury. The situation is actually directly analogous to (though more complex than) the problem of concomitant H_2 evolution during metal extraction, as discussed in Section 5.3. You saw there that the relative rates of two competing processes can be estimated by comparing their experimental Tafel plots. In this case, we were unable to find Tafel data for the desired process (equation 53), so a detailed analysis of this sort is not possible.

(c) Equation 54 represents the reaction between sodium amalgam and water, which is apparently kinetically hindered *within* the cell, but occurs rapidly *outside* it – in the denuder. A plausible explanation is that the transition metal (Ni or Fe) catalyst provides an *alternative* surface – alternative to mercury, that is – for the H_2 evolution reaction. (Certainly i_e is much higher, about $10^{-1}\ \text{A m}^{-2}$, on these metals.) This is seen most clearly if you think of reaction 54 as a sort of 'short-circuited' electrochemical process:

$$2Na(Hg) \longrightarrow 2Na^+(aq) + 2Hg(l) + 2e$$

and

$$2H_2O(l) + 2e \xrightarrow{Fe/Ni} H_2(g) + 2OH^-(aq)$$

where the two electron transfer processes occur on different parts of the impregnated graphite surface. As you will see in Topic Study 3, metallic corrosion is thought to take place by just this type of mechanism. (In addition, neither iron nor nickel readily forms an amalgam, so there is no risk of losing the catalyst.)

(d) Provided the anode-mercury gap is as narrow as possible, the lack of a separator should help to reduce the cell resistance: in practice, it does – as does running the cell at 60 °C. (Raising the temperature will also affect the emf of the cell, of course – and improve the electrode kinetics.)

ACKNOWLEDGEMENTS

Grateful acknowledgement is made to the following sources for material used in this Block:
Figures 35, 38 and 41: D. Pletcher and F. Walsh, *Industrial Electrochemistry*, Second edition (1990), Blackie Academic and Professional; *Figures 31 and 36* D.L. Caldwell in Bockris *et al.* (eds) *Comprehensive Treatise of Electrochemistry*, vol 2, Plenum Publishing (1981).